A MACHINE
TO MAKE
A FUTURE

A MACHINE
TO MAKE
A FUTURE

Biotech Chronicles

Paul Rabinow

and

Talia Dan-Cohen

PRINCETON UNIVERSITY PRESS
PRINCETON AND OXFORD

Published by Princeton University Press, 41 William Street, Princeton,
New Jersey 08540
In the United Kingdom: Princeton University Press, 3 Market Place,
Woodstock, Oxfordshire OX20 1SY

Library of Congress Cataloging-in-Publication Data
Rabinow, Paul.
    A machine to make a future: biotech chronicles / Paul Rabinow and Talia Dan-Cohen.
      p.   cm.
    ISBN 0-691-12050-1 (cl.: alk. paper)
    1. Celera Diagnostics.   2. Human Genome Project.   3. Biotechnology
industries—California.   4. Business anthropology.   I. Dan-Cohen, Talia, 1982–
II. Title.
QH445.2.R336 2005
338.7'6606'09794—dc22   2004045778

British Library Cataloging-in-Publication Data is available

This book has been composed in Minion

Printed on acid-free paper. ∞

www.pup.princeton.edu

Printed in the United States of America

10 9 8 7 6 5 4 3 2 1

To the families Dan-Cohen
and Rabinow—
past, present, and future.

Contents

Overture

A MACHINE
TO MAKE
A FUTURE

In 1999 there was unanimity of opinion within the community of molecular biologists that the number of genes in the human genome was around 100,000.[1] Scientists we questioned before the completion of the genome project dismissed queries as to whether this figure could be substantially wrong. When the human genome was at last sequenced, however—a monumental technological achievement—the figure came down substantially. The announcement, in 2000, that there appeared to be fewer than 30,000 came as a surprise. Many scientists immediately claimed that the numbers were not biologically significant. More reflective scientists are already engaged in debates over how best to define a "gene" or how to understand, as it is increasingly phrased, "gene action."[2] This is just one example of how the completion of the sequence of the human genome has brought new questions to the fore and unmasked the vulnerability of old assumptions. Many things are at stake in this emergent work: understanding of basic life processes; new approaches to disease, diagnosis, and treatment; large sums of money; technology development; self-understandings of health and disease; and so forth. In these domains of life and the life sciences, a future different from the present—perhaps radically so—is in the making.

Celera Diagnostics (CDx)—the biotech company in Alameda, California, whose formative work during the year 2003 constitutes the subject matter of this book—has wagered that the knowledge becoming available about the human genome and its implications for human health can now be turned into a powerful diagnostic apparatus that will continue to produce an incessant proliferation of new details while yielding few surprises of a magnitude that would put their strategy fundamentally in question. Of course, there will be refinements, modifications, and updates, but the core insights about how the genome functions to cause or affect diseases will retain their pertinence for a number of years. Celera Diagnostics, we venture to say, seeks to construct a "machine to make a future." One aspect of life in the future, the identification of health risks through genetic testing, will be—if Celera succeeds—predictable, conventional, and medically relevant, largely anonymous and widely distributed. It will form part of the fabric of how health, and health care, is understood, practiced, and managed.

Anonymity, however, does not imply homogeneity or stasis. The apparatus that standardizes knowledge in genetics is perpetually under construction. The people, places, and techniques involved in this fledgling apparatus are far from stable. It remains unclear whether knowledge and technology can be brought together in such a way as to be productive in the field of genetic diagnostic tests. As we are assembling this chronicle during the summer of 2003, it is not possible for us to know what corporate successes or failures might mean for the composition of the machine, and hence for the shaping of a future. But the larger undertaking of scientific discovery, the one that ceaselessly introduces new questions, and welcomes and requires only temporary stabilization, is neither feasible nor desirable for a company like Celera Diagnostics. Celera Diagnostics is in one sense involved in a far more mundane enterprise: it aims to become a routine part of the way medicine is done while continuing to innovate. Should this enterprise succeed, it would no doubt be an important machine, one that would make a future. It would be a machine, however, that would certainly lead to other such machines. And those machines, in turn, would play a role in shaping, or at least inflecting, aspects of yet another future. Thus we can say that Celera Diagnostics is attempting to build a machine to make *a* future but not *the* future.

A Machine to Make a Future

The French biologist François Jacob, winner of a Nobel Prize, writes in his book *Time and the Invention of the Future*, "What we can guess today will not be realized. Change is bound to occur anyway, but the future will be different from what we believe. This is especially true in science. The search for knowledge is an endless process and one can never tell how it is going to turn out. Unpredictability is in the nature of the scientific enterprise. If what is to be found is really new, then it is by definition unknown in advance. There is no way of telling where a particular line of research will lead."[3] There is something wonderful about this image of researchers struggling to define questions for as yet unknown answers, living amidst the unpredictable while being open and attentive to the unknown.

Such a state, however, is not a contemplative one, nor is it carried out without material and conceptual restrictions. Jacob partially makes this point when he writes in his autobiography, *The Statue Within*, "In analyzing a problem, the biologist is constrained to focus on a fragment of reality, on a piece of the universe which he arbitrarily isolates to define certain of its parameters. In biology, any study begins with a 'system.' On this choice depend the experimenter's freedom to maneuver, the nature of the questions he is free to ask, and even the type of answer he can obtain."[4] The graceful flow of Jacob's prose is captivating, but it glides smoothly past several ambiguities that are worth pausing over. For example, Jacob links three terms—"a problem," "a fragment of reality," and "a system"—in an untroubled manner. Upon reflection, one would like to hear more about how such problems are historically shaped or inflected, and by factors more diverse, more down-to-earth, than scientific curiosity alone; how conceptually difficult it is to establish relationships between scientific problems and fragments of reality; how technological constraints condition the emergence and articulation of the system Jacob invokes; and, finally, how local particularities are transformed to arrive at claims and procedures applicable elsewhere.

The German historian of science Hans-Jorg Rheinberger has devoted much erudite effort to elucidating these topics. Although his exposition is far too technical to present in detail here, the following claim is pertinent: "Experimental systems are to be seen as the smallest integral working units of research. As such, they are systems of manipulation designed to give un-known answers to questions that the experimenters themselves are not able clearly to ask. Such setups, Jacob once put it, are 'machines for making the future.' They are not simply experimental devices that generate an-swers: experimental systems are vehicles for materializing questions."[5] Again, such prose carries us along too quickly, or perhaps it refers to highly distinctive situations. We have no reason to doubt that the kind of scien-tific "machines for making the future" described by Jacob and Rheinberger did exist; perhaps they still do. However, we also have reason to believe that there are other types of machines that make futures.

Just as there are machines built to operate in a sea of uncertainty, there are others whose goals and whose operators live in a contrastive environ-ment. Once they appear, other machines are invented to stabilize them and make them available in a standardized, consistent mode, one that can be transported from context to context, or, more accurately, brings with it a context of its own. Making "immutable mobiles," to use the telling phrase of Bruno Latour, is hard work. Frequently, it is not especially glorious work. But contemporary society would not exist if there were not a vast number and variety of such machines functioning to produce movable, standardizing operations.[6]

About the Book

This book presents aspects of diverse developments taking place at Celera Diagnostics, largely during the year 2003, without seeking to force these events, technologies, scientific findings, business pressures, and interper-sonal dynamics into a unified historical narrative. By unified we understand a story with a beginning, middle, and end. The narrative form has domi-nated almost all nonfiction writing for the last two centuries. Although this genre fulfills the reader's expectations of how a plot line should proceed, it nonetheless carries with it some disadvantages. These disadvantages lie pri-marily in unearned claims of opening and closure. Thus, we have decided to experiment with a form that we hope will be more open, more tentative, and to that degree closer to how things happened.

To our knowledge, all books written by journalists—and there are many—that treat one or another aspect of the human genome mapping project or the biotechnology industry have been cast within the mold of the traditional historical narrative. Many of these books are informed and informative, providing valuable documentary evidence and insightful commentary. They tell a story of vivid actors, dramatic events, and significant scientific discoveries; they include accounts of combat, of winners and losers, of startling results. They usually end with a statement about how the future will be different. The much smaller number of academic books treating the genome mapping projects or the biotechnology industry employ a similar form. There are, of course, differences between scientific journalism and academic writing: most commonly, the academic narrative opens with reference to an ongoing debate within the scholarly literature; it then proceeds to explore its case study in light of that debate and its associated theories, arguments, and polemics.

In recent years, these two subgenres, science journalism and science studies, have told their readers stories that depict apparently quite different worlds. In science journalism, we encounter a world of passion and achievement; in science studies, a world of power, profit, and illusion. The dissonance between these depictions is striking and initially puzzling; the puzzlement recedes, however, once one realizes that these are not so much disparate worlds being represented as clashing claims set within a similar genre. The forms are similar although truth claims, ethical implications, and social consequences are evaluated quite differently. In what follows, the reader will encounter no heroes and heroines, no evildoers, saviors, or exotic personalities; nor do we offer a well-crafted account of an epoch transformed for good or evil. Rather, we seek to present the vicissitudes, strategies, and tactics of an emergent project during a finite period of time, as articulated by the actors themselves.

We decided to work primarily through a set of lengthy interviews. In part, we chose this form for a rather traditional anthropological reason: to provide ample space for what has been called "the native's point of view."[7] We were interested in seeing what would happen if we gave substantially more narrative space to the people and events at Celera Diagnostics than they would usually receive from either journalists or scholars. Additionally, we decided to punctuate the book with interviews from actors familiar with Celera Diagnostics but viewing it from an adjacent position. These interviews are intended to play a number of roles. They are designed, in part, to create an anthropological archive of a biotechnological event (most traces of which have already disappeared, as have the traces of countless others) for others to use later, and as suits their purposes.

One core reason the research, and especially the interviewing, could proceed so rapidly by anthropological standards is that Rabinow had been working with many of the key actors for well over a decade; he had previously written a book, *Making PCR: A Story of Biotechnology*, to which many of the same scientists had contributed their time and insights. This previous familiarity, confidence, and trust meant that many of the informants were already substantially comfortable with what this form of anthropological interview sought in terms of content, detail, and even tone. Further, all of these informants are self-reflective people involved in the early stages of a project that required incessant discussion and self-evaluation as well as formal and informal presentations both to others within the company and to potential collaborators in the academy or industry. Finally, as opposed to many anthropological projects, here there was no problem of access to informants and no language barrier. Of course, more interviews were undertaken than appear in the book, and not everything covered in the selected interviews has been included. The choice of what to include and what to exclude turned largely on issues of narrative flow, overlap of scientific content, and the elusive quality of "voice." Hence these choices turned as much on stylistic judgments as on some phantom methodology.

The form also serves, albeit to a more limited degree and in a backgrounded manner, to allow the anthropologists to place themselves within the chronicle. Although this presence is kept to a minimum, it would be disingenuous to pretend either that the chronicle simply unfolded by itself or that the dynamics of the interviewing process were spontaneous. The interviews contained in this book are interventions elicited, transcribed, and framed during a specific period of time. Furthermore, they are not put forward as a full description of a culture, a place, an epoch, or even an event. They are partial descriptions, chosen from a point of view, framed to manifest some aspects of a certain actuality. Our hypothesis is that by giving to our inquiry and to our narrative a form different from traditional anthropological monographs, we will produce a different kind of result, one that is not intrinsically more valuable but one that will frame a different experience for readers and writers alike. Evaluating the results is a task for the future, again one for readers and authors alike.

The other experimental dimension of our form concerns what the French thinker Michel Foucault called the "author-function." Who claims credit for a piece of work? Currently there is little ambiguity concerning this issue; it is either the professor (who thanks others for their aid and support in a section called "acknowledgments," a historically recent Anglophone invention) or the journalist. Today, and with increasing frequency, busy scientists tell "their story" with the help of scientific or financial journalists, whose aid is generously acknowledged. The rules of authorship in scientific journals are stable; there is a clear-cut set of procedures to establish priority and sequence of authorship. There is an equally clear-cut and strictly defined discursive form for scientific articles. Authorship and form have been joined together and successfully linked to a reward structure of careers. That is, the rules for attribution are linked to a specific form of writing.

In contradistinction, A Machine was produced through an unorthodox arrangement that has yielded a distinctive author-function. Paul Rabinow is a senior professor of anthropology at the University of California at Berkeley. At the time the research was conducted, Talia Dan-Cohen was an undergraduate at Berkeley. Rabinow was curious to see if he could put together a chronicle in which the authority of the author would recede as the organizing principle but, as in the sciences, the crafted and disciplined use of different perspectives would enrich the results. In the fall of 2002, Rabinow proposed such a project to Dan-Cohen, and she accepted his unexpected and challenging invitation to collaborate. The research and the writing took nine months, from January 2003 through September 2003.[8]

This relationship between a senior professor and a gifted undergraduate was not one of principal investigator and research assistant. Clearly the impetus and contacts for the project came from Rabinow. However, he had felt for some time that it was vital for an anthropological inquiry that sought to come to grips with a fast-moving actuality to develop a different research method. One dimension of this experimental method was to include another observer. The question remained, How? One way would be to do a joint project with another professor or a graduate student who was already versed in the scholarly literature and committed to an academic career, hence already socialized into existent questions, existent answers, observational and analytic skills. Although that approach can be fruitful, we chose instead to see if the structural limits of a situation could be turned into an advantage. Dan-Cohen's role was primarily, in the insightful phrase of the sociologist Niklas Luhmann, to "observe the observers observing."[9] This meant that the directed interviews between self-reflective informants engaged in formative work and an experienced anthropologist engaged in active deliberation with them were themselves subjected to observation. While Rabinow and the informants talked, Dan-Cohen observed. This structural position contributed to her fresh insights into a range of dynamics not readily discernible by people who already know each other and thus can take certain things for granted; and this led to discussions and consequent further probing in directions that the engaged anthropological observer would otherwise have been blind to by dint of the very conditions that had facilitated the work in the first place. The work of writing, and rewriting, was basically shared equally.

A Prehistory: From Celera Genomics to Celera Diagnostics

An important element in the chronicle of Celera Diagnostics is its attachment to, and reliance upon, the more glamorous Celera Genomics and its past achievements. Although Celera Diagnostics is a separate company, it has certainly taken advantage of its access to the people who led the sequencing project, the technology with which they achieved their results, and the company name under which they did so. This historical backdrop, therefore, is still important within the company, but it does not define the character and attitude of Celera Diagnostics' management strategy. Craig Venter is no longer running the show, and his replacement is unquestionably not trying to play the role of the scientific maverick.

The founding of Celera Genomics in 1998 was a major turning point in the worldwide efforts to map the human genome. The company's genome sequencing strategies, as well as its publicity machine, introduced turbulence of an unprecedented sort into the already highly competitive world of genome mapping and sequencing. The founding of Celera—"Celera" is derived from the Latin word for "swiftness"—as well as the turbulence it created and exploited has been well documented.[10] These accounts focus attention, appropriately, on the confrontation between public and private institutions and the accelerated mapping of the human genome that resulted from this competition.

When Mike Hunkapiller, president of Applied Biosystems (AB), a leading instrument company, approached Venter with the idea of creating a company that would sequence the entire human genome using AB's new PRISM 3700 instrument, Venter was ready to attempt it, and in 1998 he launched Celera Genomics. The possibility of a private company challenging the public effort was now on the table. Celera's facility would contain hundreds of AB's new PRISM 3700 sequencers, as well as an $80 million Compaq computer that would handle the enormous amounts of data that the sequencers generated daily. Celera had been founded for the purpose of deciphering the 3 billion base pairs of human DNA, and it now had the resources to succeed.

Venter had yet to prove that it was possible and sensible to use the shotgun method on genomes significantly larger than those of bacteria. To do so, he undertook the sequencing of a popular model organism, *Drosophila melanogaster*—the fruit fly. At that point, a joint effort between the Berkeley *Drosophila* genome project, headed by Jerry Rubin, and the European *Drosophila* genome project was slowly progressing in its own sequencing efforts. In 1998, Venter proposed that Celera and the joint public *Drosophila* project work together to rapidly sequence the *Drosophila* genome using the shotgun method. Celera would provide much of the funding for the collaboration. Since the public *Drosophila* genome project was less than a third of the way through after four years of work, and relied on grants, Venter's proposal for collaboration was accepted. Rubin's decision to cooperate with Venter was heavily criticized as a Faustian bargain. In written testimony before the House of Representatives, having completed the sequencing successfully, Rubin admitted that his decision to cooperate with Venter had not been supported by his colleagues: "Many colleagues were not enthusiastic about a collaboration with a for-profit company on the genome project, despite the fact that academic researchers develop partnerships with the pharmaceutical and biotechnology industry all the time. A lot of my friends were particularly leery of collaboration with Celera. They warned me that I was going to get into real trouble and would feel badly treated at the end of the day."[11] Later in his testimony, Rubin affirmed that the cooperative effort had not only sped the process of discovery but had also been an enjoyable experience: "Celera honored all the commitments they made to me in this collaboration and they have behaved with the highest standards of integrity and scientific rigor."[12] *Drosophila* sequencing was completed years ahead of schedule, and Celera proved the value of its

method and its machines. The March 24, 2000, issue of *Science* was largely
dedicated to the successful completion of the *Drosophila* genome sequence.
The results represented a distinctive achievement in genetics and biology
and raised Celera's credibility for its more grandiose sequencing goal. This
increased credibility, of course, equally intensified the fear that Celera
might succeed.

Nearly two weeks later, Celera announced that it had completed sequencing random segments segments of DNA belonging to the human genome
and was now ready to assemble the pieces. All the while, Venter continued
to defend Celera's intentions regarding how, when, and to whom the company's findings would be disclosed. Furthermore, tensions between the
public sequencing project and Celera's efforts were growing as each side
publicly emphasized the shortcomings of the other. The public human
genome project was preparing for the release of its rough draft, while Celera was speedily assembling the now sequenced segments of human DNA
for its own draft of the genome.

In the spring of 2000, the two sides were being pressured to resolve some
of their differences for the purpose of a joint announcement. The pressure
came largely from President Clinton. Realizing that the ongoing, very public bickering was undermining the dignity of the achievement, and making
both Celera and the public project look ridiculous, the leaders of the contributing parties came to a diplomatic resolution. Wade explains the form
of the resolution: "An elegant, minimalist deal emerged. Its terms seem to
have been along the following lines: The two sides would make their vic-
tory announcements together and publish descriptions of their achievements simultaneously, though in separate articles and maybe separate
journals. Neither would publicly criticize the quality of the other's work."[13]

On June 26, 2000, Venter and Francis Collins, the director of the government genome effort, came together in the White House to announce
their respective achievements and to celebrate what represented a historic
moment, a major milestone for science and humanity. Venter intoned,
"Today, June 26 in the year 2000, marks a historic point in the 100,000 year
record of humanity. We are announcing today that for the first time our
species can read the chemical letters of its genetic code."[14] These claims
were vague so as not to repeat the blunder of a March news announcement
that the knowledge of the human genome sequence belonged to humanity.
That press phrase had caused a vast hemorrhaging of investment in bio-

technology stocks. By the time Blair and Clinton backpedaled the damage was done. Celera's stock, which peaked at $247 in March 2000, was trading in the $15 range by April 2002.

Venter, the man who represented money-driven, scientific exploration to the scientific community as well as to the general public, was not committed to Celera Genomics as a business enough to restructure the company and its business model. Applera's CEO, Tony White, made it clear, however, that with or without Venter, Celera Genomics would undergo some sort of transformation. In April 2002, Celera Discovery Systems, the genetic databases used by drug companies, were transferred to the profitable Applied Biosystems. Selling subscriptions to genetic databases had been profitable in the short term, but now the publicly funded human genome project was offering for free much of the information that was sold through the Celera Discovery Systems databases and thereby undercutting Celera's business plan. In a conference call with investors, Tony White remarked, with his dry wit, "Our friends in the public sector have been more successful than we thought they were going to be. We decided to go ahead of that curve and not wait for this thing to deteriorate on us."[15] Hiring Kathy Ordoñez as president of Celera Diagnostics in October 2000, and then as Venter's replacement at Celera Genomics in April 2002, amounted to essential first steps in the recrafting of a business plan.

Kathy Ordoñez is nothing like Craig Venter. While Venter stole the spotlight with his scientific adventures, Kathy Ordoñez has remained, for the most part, under the radar of the media. Ordoñez's agenda has been far less attention grabbing, though her resume is certainly impressive in its own right. A *Washington Post* article from 2002 describes Ordoñez as follows:

[Ordoñez] is that rarity among corporate executives—female, for one thing. Knowledgeable about the mechanics of her business, for another. And shy. Ordoñez would rather strategize and hypothesize than pontificate or pose for magazine covers. [. . .] Ordoñez has five patents under her belt, a following among genetic researchers and significant marketing achievements. She has commanded teams of hundreds. She has driven others to success, often stretching them—and herself—beyond regular hours and job descriptions. She just hates being the center of any attention.[16]

Kathy Ordoñez graduated from Hartwick College in 1972 with a B.A. in chemistry. In 2000, Hartwick honored Ordoñez for her achievements with an honorary Doctor of Science degree. By that time, Ordoñez had already been president and CEO of Roche Molecular Systems for nine years. She had constructed a team whose aim it was to take on the development and commercialization of diagnostic products that relied on polymerase chain reaction (PCR) technology, thereby exploring ways in which new scientific knowledge could be turned into useful and efficient products and made profitable in industry. In December 1991, Hoffman La Roche had paid Cetus Corporation $300 million for the rights to PCR.

When Ordoñez was named president of Celera Genomics, analysts and business experts were quick to voice their surprise. Ordoñez was not the obvious choice. She had no prior experience in drug development, which constituted the primary objective in Celera Genomics' reformation, and some newspapers reported that Ordoñez was hired after a number of candidates with drug development experience had declined Tony White's offer. Tony White himself defended his choice by pointing out that Ordoñez had a solid track record in picking up new projects, was well versed in business *and* science, and was accustomed to quick decision making, something with which pharmaceutical company executives were less comfortable.[17] Since being named president of Celera Genomics, Ordoñez has been splitting her time between the Maryland and California facilities. She is reshaping Celera Genomics from the ground up, while also constructing and overseeing Celera Diagnostics.

Chapter One

ENDING
AND
BEGINNING

Confronted with an apparently deadlocked and frustrating situation, actors have a number of possible courses of action. The economist Albert Hirschman developed an elegant typology of such options.[1] The first is one of "voice": the actors remain in the troubled situation but actively seek to propose alternatives. By so doing, they affirm their fundamental loyalty to the current order of things but express their dissatisfaction with it. By affirming their loyalty, they legitimate their criticisms as being in the interest of the organization, product, or party. A second alternative is to remain loyal to the organization, product, or political party and simply endure the strain of the current situation, hoping that it will change. The third option is "exit." This option usually is chosen only after attempts at voice and loyalty have failed. Exit can be a kind of voice, as it makes a statement of an informed kind, addressed to those who have the power to change things, about what the actors take to be an untenable state of affairs. Exit can even be a kind of loyalty, in the sense that it may well affirm commitment to the fundamental principles or modes of operation that moved the actors to join the organization, buy the product, or work for the political party in the first place.

The future directors of Celera Diagnostics proceeded haltingly and with much reflective, even agonized, soul-searching, down the paths of voice, loyalty, and finally, exit. In some situations, stasis and patience may well be a plausible option. One can imagine political loyalists waiting through a rough period until better times arrive, but in the domain of genomics there is no such thing as long-term stasis. On the one hand, as the investments, stakes, and pressures are so high, decisions need to be carefully weighed, especially as they are being taken in uncharted waters. (No one knows how to best do genomics.) On the other hand, the investments, stakes, and pressures preclude excessive delay or procrastination. Everything turns, therefore, on what one considers to be "excessive."

In addition to differing judgments as to what constitutes a "reasonable" or "prudent" weighing of options—over which actors might well differ in good faith—there are other factors that affect the dynamics of loyalty, voice, and exit. These include the legitimacy accorded to those making decisions. That legitimacy rests on multiple foundations, from a simple respect for hierarchy to a sense that even decisions one disagrees with are nonetheless being taken in good faith and are being applied in a spirit of equity. When the latter affective or emotional ties are eroded, they become very hard to repair. Honest differences of opinion are easier to adjudicate and overcome than the erosion of trust. Once the latter process advances, the proverbial exit door beckons. Of course, upon exiting one must go somewhere else. Hence conditional planning and increasingly divided loyalties (even if the letter of the law is strictly followed) almost invariably precede leave-taking.

The leaders of Roche Molecular Systems—Kathy Ordoñez, Tom White, and John Sninsky—during 1999 and the first half of 2000 found themselves in a situation that fits Hirschman's typology. To show them facing the fork in the road, we first present a summary of the situation as seen from the outside, including an interview with Michael Hunkapiller, president of Applied BioSystems, and then, two interviews with Tom White, as well as interviews with Kathy Ordoñez and Gabriella Dalisay, White's executive assistant.

Holding Pattern: Turbulence and Stasis

A line of developments with direct consequences for our chronicle took place in 1998 when Hoffmann La Roche acquired a German company, Boehringer-Mannheim. The merger was announced in April 1998 and officially started in April (or May) 1999. Mergers are complex affairs requiring multiple levels of negotiation to bring about and many more steps to bring to fruition. Among the possible complications, and in this case an apparently unanticipated one, is cultural friction between the merging companies: German, Swiss, and American styles of management and personal relations did not blend easily. This cultural friction was linked to the power dynamics any such situation entails. The German representatives jockeyed hard for increased authority and power in shaping and running the new organization, and over time they were making advances in achieving their goal. There is no need to explore here the typical corporate and bureaucratic politics involved, only to underscore that they existed, and that they set other things in motion that do concern us directly. Among these factors was the interpersonal and intercultural dynamics between a company run in an American style and headed by a woman and a more bureaucratically oriented company staffed in its upper echelons by men of a certain age and style.

Thus, for the key players at Roche Molecular Systems, it was during 2000 that future career options became an object to reflect on. Equally, as we have seen, it was during this period that the developments in genome mapping were coming to fruition as the race to finish the initial sequencing accelerated.

We interviewed Mike Hunkapiller at Applied Biosystems on July 7, 2003. He explained the events that led to the departure of Venter, in January 2002, from the company that he had helped make famous and the arrival of Kathy Ordoñez onto the scene.

MH: Celera Genomics was started as an information and bioinformatics endeavor, and its initial goal was to use the sequencing of the human genome to establish a position in that field. It was not just sequencing for sequencing's sake, and it absolutely wasn't to build a huge patent portfolio around the human genome as such. It really was to lay the foundation for an information business, and from my perspective, it got a little bit ahead of itself as being seen as a business built around a proprietary set of information, and it was always intended to be an informatics tools business. The value lay in helping people understand what the human genome was. Celera Genomics had envisioned, even early on, taking some of the information tools applied to the sequence data and pulling out bits of information that they would exploit themselves, either as therapeutic targets or as diagnostic targets.

They didn't have the expertise built up to do either because things moved so fast on the sequencing side that the company got ahead of itself.

PR: So Celera's strategy of challenging the public effort worked almost too well?

MH: Almost too well, that's true. They chose to focus initially on the therapeutic side. And partly because that's a long-term research endeavor you do internally, we decided that the best way to do the diagnostics was to create a joint venture between two of us, and that provided a good vehicle for bringing Kathy and her group in.

PR: So there was no resistance from Celera Genomics on any of this?

MH: No, I think Craig would have probably preferred to do a lot of it in Celera Genomics, but he really didn't have the people and the expertise to do it, and if you're going to bring on somebody of Kathy's caliber, they have to have a coequal position in the overall management scheme for the purpose of arguing for resources. So the resistance didn't last for long.

PR: She speaks only glowingly of Venter, which is what one would expect.

MH: [*laughs*] Well, I mean, Craig has had a history of successfully challenging conventional wisdom as to what's possible scientifically. And he likes to play the role of a maverick in that process, there's no doubt about that; he's been right on many occasions when the traditional wisdom wasn't so correct. While Celera was in the formative mode of really having to challenge what was thought to be the pathway to get the human genome sequenced, Craig was really into that. Once he had to step back into a more traditional, longer-term role of managing a business, he was less well matched for the job and the role in the therapeutics aspect. Being a maverick may sound good but it doesn't work with the FDA or the whole procedure of getting things through a long, drawn-out process. And so he was just less comfortable with doing that. I think it was a natural parting point. Had things moved more slowly in the sequencing, he would have had the time to build up an organization that would have brought in the relevant expertise to take that over, but there wasn't time.

PR: And therapeutics is sexier than diagnostics too?

MH: Well, I'm not sure I would argue that that's the case—some people might. It has bigger payoffs associated with it in some cases, but I wouldn't argue it's any more valuable or interesting scientifically. But Craig hadn't done the diagnostics either. He just hadn't built the biology up commensurate with the science associated with generating large amounts of information, which is what the expertise of Celera Genomics was initially—bioinformatics. And they had begun to do a little bit on the protein side, mostly from the perspective of coming up with protein targets for therapeutics. But it was still research, not research directed toward specific disease indications. It was kind of broad: How do you generate a lot of data and then begin to pick the cream off the top of that? So when he left, Tony [White] looked to bring in a seasoned pharmaceutical executive who could mesh the research endeavor and engine that was there with the opportunities in therapeutics. In the end he decided that maybe that wasn't the best position for us to be in, because it's a rough road—it may be sexy, but it's also risky, and the failure rate is very high. Is there a way that we can take advantage of the fact that we've got a pretty successful beginning to a genetic diagnostics business and the right team there in Alameda running it? So we decided to try doing the clinical diagnostics and the clinical therapeutics development together in areas where there could be synergy between the two.

PR: So it was very intense?

MH: Sure. Celera had taken advantage of the big spike in market capitalizations to go out and do a secondary offering and raise money. We had the resources there. We didn't want the resources sitting idly.

PR: And then, given that thinking, it was, as you say, natural that Kathy and her people would be a good choice?

MH: Well, Kathy was brought in to run the diagnostics business before Craig left. Although they're separate entities, having Kathy oversee both of them could maximize the synergy between the two.

Initial Plans: Interview with Tom White, October 15, 2001

TW: Today is the anniversary of the date I accepted the job offer from Applied Biosystems. The same applies to Kathy and John. Four to six weeks later we left Roche and joined Applied Biosystems.

The first few months were spent analyzing the AB technologies; we were somewhat familiar with them as AB was Roche's partner in the research field. We spent close to three months working together by driving down to Foster City a lot. We would meet in the morning at Kathy's house from eight to ten until the traffic died down, then head across the bridge and try to leave there by three o'clock, when the traffic got impossible. So it was a chance to look at things without the encumbrance of having the Roche business going. We were essentially developing a strategy for analyzing what we were going to do. At Roche we were so busy we were not able to devote the extensive time to do this analysis.

We began with a clean slate about AB and then developed a plan about what to analyze. The existing molecular diagnostic business in infectious disease tests is Roche's business, and that is the one we built; we know the competitors (Abbott, Bayer). They are still thinking about how to take Roche's business away from them today rather than thinking about how to proceed over the next ten years. They have had trouble competing against Roche, rather than leaping into a completely unknown area. This is not a good strategy since Roche's competitors had not done well in the infectious disease area. Other companies approached us because they had been competing against us at Roche, and they were now thinking it might be useful to work with us as a way to compete against Roche. For us it was more a question first of figuring out what we wanted to do. We didn't want any more encumbrances; we felt that those companies' limited success in this field was not a plus. We left to do something really different. We are focusing on molecular diagnostics, a totally new field that is still only a small part of the diagnostics field as a whole, which is mostly clinical chemistry. Molecular methods have increased to about a billion dollars a year over the last ten years, but that is only 5 percent of the diagnostic product market. There are only four or five companies that are in that. Since those companies also offer the full range of diagnostic tests, a new, small company can't really compete in hospitals, medical centers, and big labs unless it offers the full range of products. That is the flaw of most of the biotech companies; they think, "We have hot new technology or a hot new test for a specific thing," but they don't realize how diagnostics are delivered to the worldwide medical system. Even when we were at

Roche, we had R&D, manufacturing, regulatory, et cetera, but we did not actually sell to customers; we sold to another unit of Roche. We wanted to discover what was useful medically, develop it, and then have someone else sell it. And then we can go on to the next medically important thing. We felt that we don't want to get into the end stage business of marketing and sales to customers. This meant we had to partner with one of the big companies like Roche or Bayer or Abbott, BioMerieux, GenProbe, or Johnson & Johnson. It was always an element of our strategy to figure out who was the best company to distribute the things that we develop. Because of our past relationship with Roche, our former employers, we were not inclined in that direction. At the same time, they own more than half of the molecular diagnostic business and they are more than five times more powerful than any of the other second choices. We wanted to explore the thoughts of the other companies; by the end of the year 2001 we intended to finish our discussions. So that is the strategy of establishing the downstream part of the business that will allow us to feel more comfortable about focusing on what we are calling "the front part."

The Front Part

We began by thinking about what were the areas that Roche had not analyzed. Roche has described their genetic projects in a number of areas, such as heart disease and certain inflammatory diseases that are tied to Roche's pharmaceutical interests. Since founding Celera Diagnostics, we have analyzed over two hundred complex diseases for unmet diagnostic needs. It is the most comprehensive analysis I have ever seen. To my knowledge, no other diagnostics company anywhere has done this kind of analysis. The quality of the analysis is pretty amazing in terms of what we have selected to do. So there is no reason to not just forge ahead and do it. We are not like Incyte, Genaissance, DNA Sciences, or little biotech companies trying to raise money by claiming they can do everything. We are prepared to methodically set up this massive thing to do disease association studies on the same scale as Celera Genomics sequenced the genome.

There have been a couple of illuminating moments. One was when it appeared in February 2001 that the total number of genes was on the order of 30,000. Whereas before we thought we could study panels of candidate genes, 500–1,000 genes known to be useful in heart disease or lung cancer, it became clear that we could probably study the whole genome and not be limited to panels. With a different approach and a thorough calculation of the costs, we could study everything. And once we had that set of SNPs in place for the whole genome, there was no reason in the future to study only the smaller subsets of candidate genes. Kathy, John, and I had the idea at the same time, and we calculated our costs based on using the technologies at our disposal. We realized that no other company could do what we intended to do, because of this unique combination of technology that we could bring to it. It was the scale of the whole thing. We will always want to study a thousand patients for statistical power: five hundred cases and five hundred controls. This would give us the statistical power to find associations, if they exist. Then we will obtain genotypes on one thousand patients using ten, twenty, or thirty thousand SNPs. Can we do that at a reasonable cost, in a reasonable amount of time? Yes.

We then looked at every other group that was trying to do this (whether it was the public project with other SNP databases, or Sequenom or Incyte, etc.). They lacked the scale. Thirty thousand SNPs in 1,000 patients—what would it take? We were so used to thinking about budgets and their constraints. Kathy said, "Don't think that way—think what it took to set up Venter to do the genome project; think about a lab with one hundred instruments, working twenty-four hours a day. How long would it take to analyze all diseases?" We did the calculations and decided, "This is what it would take."

We have been trying to describe the workflow process and only recently figured out a way to present visually what we are trying to do. What was the conceptual approach? So much of what's been done is linkage and linkage disequilibrium—having a picket fence of markers and going across the genome. Well-placed pharmaceutical people were saying that you need 200,000 markers, and soon everyone will go to the doctor and these will be measured; that will tell everything that we need to know about health risks. But such a claim ignores the reality of how diagnostics works on a practical scale, and doesn't seem to understand the statistics involved in studying 200,000 SNPs. The

trouble is that it would cost almost $100 million. We are going to con-
centrate on the genes and not on the other 99 percent of the genome.

PR: What is your key question?

TW: Can it be done? It must be cheaper than other companies' approaches.
The steps are expensive; we know what the key reagents cost for most
companies; we can calculate that they simply cannot be doing those
studies for less than tens of millions of dollars or more. There is no
way they can raise that kind of money to do this kind of study. We
know our costs are lower, because of our alliance with Applied Biosystems. It comes down to scale and cost at that scale; only the big
pharma companies could do this. We have no competition as far as we
know.

Strategy: Interview with Tom White, March 28, 2003

PR: Let me try to characterize the situation. I am trying to identify the
major factors during the year 2000. Roche was the leader in developing the molecular technology that was used in diagnostics whose targets were known. The next stage would be to move to targets that were
not known. The other factor was that the human genome mapping
projects were now in high gear. This would make available a radically
different quantity of data to work with and set up the challenge of developing a more complicated strategy of using molecular tools to do
diagnostics.

TW: Roche Molecular Systems started in 1991 with enzymes sold through
Perkin Elmer. There were no PCR-based diagnostic products, although there were some services from Roche Biomedical Laboratories, now LabCorp. Between 1991 through the middle of 2000, the
PCR-based diagnostic product business grew to $500 million just to
Roche. The issue was to see the potential there and then to look out to
see what it could do in blood screening. Roche had only been working
on the genetics side on a smaller scale. We thought we would approach it on the scale of a few hundred genes picked on biological
grounds, because the genome had not been sequenced at that point.
You could explore the genetics in more diseases on more people. We
were essentially proposing a broader scale that one would query. It

was not what we ended up doing when we came here. Once we heard Venter's talk in Rockville, in February of 2001, we made the jump from 300 genes to 30,000 genes. There was a change of scope.

PR: In the fall of 2000, you were proceeding apace with your project but if either (a) Celera Genomics had not succeeded or (b) there were 100,000 genes or (c) Celera Genomics had not raised the large amount of money, things would have been different. The project would have been an incremental advance in the scope.

TW: Well, in 2000 we still did not know what the most important thing to work on was. We were simply too busy to do it at Roche. We never even got to the point of doing an analysis. There was no encouragement.

PR: On the other hand, no one else in the world was really doing it either.

TW: I don't think so. Even today what really counts is the question. So between April and June of 2001 (finishing up in September) we did a gigantic analysis of two hundred diseases, ten different questions. They went into the business plan in October that was approved in November, and that is what we are actually doing. The question of what to work on was as important as the genome data or the scale of the operations. The answer we decided on forms the strategy. It oriented us to get access to the relevant sample collections. There are not that many of them.

PR: Had you not been mired in the mess at Roche and someone gave you one hundred million dollars at that time, you would not be where you are today? You would not be associated with Celera?

TW: That is actually a key point. Since we had an existing business, we were totally occupied. Here, there are other ideas of equal importance to those we have already been dealing with. We are meeting the FDA at the beginning of May to show them how we could develop diagnostic products that predict adverse events that could meet their criteria for registration. These concepts have not been clearly described before. Once we left, there were only three of us. And we spent the first few months just thinking: what is it we are going to do? This was scary. The next most important thing was to think about who we wanted to attract to this new company. We looked across the whole industry for people we wanted to work with, the smartest people we could find, who were frustrated with their own organizations, who would jump at the chance to work on the genetics of common diseases, who would bring with them a wealth of experience but also know what to avoid that caused their own organizations to sink into a cumbersome bureaucracy that prevented them from acting. It was all built from scratch.

PR: What if there had been 150,000 genes?

TW: Then we would have taken a more gradual approach. In the meeting with Venter, everything changed.

PR: The genome works this way and not that way. You seized an opportunity. But if the reality was not there, one would have been forced to go somewhere else?

TW: Yes. In early 2001, people were still thinking about the genome as if it were gene by gene. They were not asking, "Why would you sequence the genome if you were going to ignore it? Go where the genes are." But then when you looked at the genes, there was not enough variation. So the next crucial thing was in the May-June 2001 period we realized we are going to have to resequence all the genes. There were not enough SNPs in genes that came out of the SNP consortium. We suddenly realized, "This is going to cost us eighty million bucks." In June of 2001, Kathy, Hunkapiller, and Venter convinced the company to spend the money. It became a $100 million investment to resequence the genes of forty people to find these kinds of SNPs. This project was way ahead of its time. We were at least eighteen months ahead of the whole field, and now we have it.

Money is a critical factor. Celera had raised money in March of 2000, largely at the peak of the boom, which basically ended when Clinton and Blair said, "No patenting." It had gone down a good deal by the time we had gotten there, but nonetheless there was a lot of money. We would have raised money by ourselves, but we would be on the verge of going out of business, like everyone else today. Smaller labs came to realize that their model was not going to work. We were looking for collaborators who would work with us jointly, and this was attractive because we could do things that they could never do. This was big science. The Celera name helped us a lot in this. There was a downside to the name, but here it helped. It helped us get established in this new game. The next stage will be the crucial one: finding the markers.

The Family of Science:
Interview with Gabriella Dalisay, June 23, 2003

Gabriella Dalisay is Tom White's executive assistant.

PR: One of our goals for this book is to let the people speak who actually make the day-to-day things run in this company.

GD: Oh, that's funny, because I've read your book before on PCR technology, and then I went ahead and reviewed it again this weekend, thinking, "What will he be asking? What is he interested in?" The way you wrote it, the people you spoke to brought the book together, and I liked that because it wasn't just your ideas; it was the group's together. So I said, "Okay, I kind of get the direction."

PR: How did you get here?

GD: Okay. I think it's kind of interesting how I ended up in biotech, because I was raised in California in a very strong Catholic family, which is pretty common among Hispanic families. I attended Catholic schools and had that education in which everything was pushed on you—the Bible, the way God created the world, the way everything evolved. When I was in seventh, eighth grade, the teachers had a brief discussion on Darwinian theory, and I thought, "Oh, that's interesting—why would they bring that up? That's so contradictory to what we grew up with. Someone just made that up." And once I resumed that study in college, and I'm still going to college (because I'll probably go to college until the day I die), but when I went to college and I learned more about it, I found that I became even more confused. How could this be? And I became more interested in Darwinian theory, but what it actually did was split my beliefs. Suddenly it was like, well, he's starting to bring me to believe what he's saying, but then I still study the Bible, so that still kind of splits me in between the two theories. Our daily living was based on the Bible, and it wasn't until I got older and I studied Darwinian theory that I became very confused. I thought, "Wow! This can't be true! This is contradicting what I grew up with."

PR: The religious crisis: did that get resolved?

GD: No. I could honestly say that I've learned to accept both, but I'm not very opinionated on either because I know there's a lot of controversy with the church. It's almost like a political statement, you know; everyone has their own opinion of it. And I guess, being raised in a strong Catholic family, I've always respected my mom and dad's ideas. I don't want to be disrespectful of the way they raised me. But being in this field for twenty-five years, I have my own ideas, and I just don't share that part. I guess I don't want to upset my parents or upset the strong family beliefs.

PR: So your parents let you be?

GD: They would question some things. I had nine siblings. I'll never forget, when I would come home from Cetus, wow! The studies! I actually saw the rats, the mice. I remember there were even studies on dogs, and my siblings thought it was so inhumane! They asked me why I'd want to be affiliated with that. I'm the oldest in the family, and I was in my twenties. So to them, the studies were being done on pet dogs, you know? There was some naïveté to that. I remember actually defending myself one time, "Well, what do you want to do? Do it on the humans and hurt humans?"

PR: These are the arguments taking place in society; they're not resolved.

GD: Yeah, exactly, even to this day. But what's interesting is that it made me think, and it kind of touched me emotionally that my siblings would feel that way, because I always had respect for family even though you'd go to college and you'd learn different ideas. I would try to challenge them, but then I could tell when it was time to stop. To this day, even my mom knows the importance of clinical trials. She suffers from rheumatoid arthritis. There was a study at UCSF that she participated in, and she saw the importance of it. It seemed to me like, "Wow, over the years, how things have changed."

I remember when I first told my parents that I was going to work for a biotech company. I told them the study I'll be involved in is about disease, people, their well-being. I saw it in a fuller perspective than my parents, and once I started working in it, I became so enthusiastic about what science can really do. They were trying to take care of people. [After an initial job] I remember there was an opening at Cetus, and I thought, "Wow, this sounds even more interesting." Cetus was involved in a wider field of study. My first job was in development. Another administrator who worked for Tom used to say, "Tom's the best. Tom is this! Tom is that!" I didn't really know Tom that well. I kind of backed her up when she was gone, and whenever I did, he was so appreciative of anything the administrator did. I thought, "Wow, this guy is great! She's spoiled! She's lucky. I mean, he actually takes the time to appreciate what everyone does."

I spent eighteen years at Cetus/Roche and during that time period, I went through five reductions in force, so it got to be very nerve-wracking. You know, every year I thought, "Oh, my God! Another one! Oh!" I couldn't stand that tension anymore. Plus, I had two kids, and I wanted more security. So when the time came, and the boss I was working for was leaving, it was time for me to move on too. At that point, I was doing project management work, and I took some courses at UC Berkeley, and I thought, "This is kind of interesting." But the more I got into it, the less exciting I found it, because it was focused on one project, not various projects.

PR: So after the Cetus breakup, you went to Chiron?

GD: Yeah, and I stayed there from 1997 to 1998, when I heard of this opportunity at Roche. I started working with Kathy Ordoñez, and I eventually transitioned into working with Tom, who was the senior vice president of R&D. This was the opportunity that I wanted. I thought, "This was meant to be." It turned out to be everything I had hoped it would be. It's not like you're just there to assist Tom when he needs you. He's a true believer in communication, and the more information I have, the more I can help him. And that's what we both learned working with each other. I do feel more involved with the science, and I've also learned that the higher the level of Tom's job—I've seen him as a director, senior director, VP, senior VP, now chief scientific officer—the more responsibility he had to take on, and the more I need to help him, because he's only one person. His workload gets bigger and bigger, and he needs someone who really can take a chunk of that. There are a lot of times I ask Tom, "What can you delegate? What can I do? What can I help you with?" And he has never been reluctant to hand over some work. He's always doing something, and it's something exciting, and that's the thing I like about it. I love a challenge. I love doing something I've never done before, and if there is something I don't know, I'll truly find a way to learn it. I think he's comfortable delegating because he knows I'll find a way to get it done using available resources. I figure you can always find a resource to help you resolve it, so that's how I get through most of my challenges.

PR: Okay. I'm often criticized for painting too positive a picture of people like Tom and this kind of organization. One question that interests me is whether you've encountered any discrimination. Or do you think that organizations like Cetus or Celera are better than some of the rest of the industry in terms of hiring and how they treat people?

GD: Personally, I don't think that's ever been an issue. As a matter of fact, the one thing I've really enjoyed for all the years I've worked in this field— with Cetus, and that was for many years, and now here—they actually appreciate the diversity of their groups. I feel like I fit right in. In fact, the diversity is not just of race but also of education, background, a lot of things. I think they see all of that as a contributing factor to the company. And I'm not just being optimistic; I'm being real. That's how I've observed it in the twenty-three years that I've been here.

PR: So you started at Roche in '98?

GD: I started in '98 and worked until April of 2001. That was interesting. It was interesting because I learned what it's like to work for a company when I saw what it did—how do I say this in a sensitive way?

PR: Do you mean sexism?

GD: Thank you! You're making it easy. The sexism is what I really struggled with. We had a female president of the company, and when I worked with her intermittently, I saw it. I feel that certain people in the new management lacked respect for her and her role as president, and that made me nervous. That made me kind of wary of where things were going to go, if she wasn't getting the support that she needed and she was running the company and we're all depending on her to make these de- cisions. But some of these people were stomping on her ideas, and pretty much, I felt, dictating how it was going to go. I thought, "How can she run a company effectively and efficiently when they're not al- lowing her to do that?" I will be honest: that made me nervous to the point where I said, "This is just a matter of time." I developed loyalty to them [Tom and Kathy] to the point where I'd stay with them and help them as long as they needed me to stay there. I wasn't going to leave during a critical time period when they needed someone there to help them. I could have jumped ship and said, "I'm getting out, because this is getting scary." Instead, I had a lot of trust in both of them, so I said, "You know what? I'm going to ride it out. It's going to be okay. These two are very smart people—Kathy, very smart business-wise. She's one of the most intellectual women I know in business. She has ideas and makes them work in a very confident way. I'm so confident it's going to work out the way she wants it to end up working out, and this may just be a bias, but with Tom on her side, it's going to happen." And the team they had working together at Roche worked very well together. Whether or not they had different ideas, they compromised, they talked. They don't all think alike, but they come together in their ideas and follow through on them, and that's what made me comfortable even during that critical time. So I thought, I'll ride it out. But I already had my mind made up that when they leave, I leave. In my mind, I knew I was going with them. I mean, I had every intention of going with them.

PR: You were in a distinctive and tough position.

GD: Oh, I'm telling you, it was the scariest time. People would look at me, ask me, "What do you know?" I couldn't disclose a thing. All that information would have created was panic and havoc and destruction for the company, so you can't disclose that. I remember going home and just saying to my husband, "There's something I've got to tell you." Having all that information would just eat me up, so I would share it with my husband, just brief points. I felt better getting that off my chest, and then I could go to work the next day and just know that it's off my chest. I didn't have to live with it on my own that way. It was a very difficult time. That's when I realized that some people didn't care about the pressure I was under. They would say, "I know you know, Gabi, tell me! Tell me what's going on!" I said, "I don't know a thing!" I said, "He wouldn't even disclose this information to me." But it got to the point where it was pretty clear that it was just a matter of time. I also knew that they wanted to handle it very professionally, very calmly. They were still sensitive to causing any disruption in the corporate functions. And I admired that, because I was thinking that if it got to the point where I couldn't even wake up and want to come to work, I'd probably check out. I'd say, "Why torture myself like this every day?" But they were thinking that, in their role, they had to be concerned for the well-being of the company, and they were.

PR: So it's a doubly tough situation for you? You're loyal to Tom and Kathy, but you had no guarantees that they were going to be able to take you with them.

GD: Oh, exactly. That's exactly what it was like. It's true. There were no guarantees. They weren't even in the position to take Roche employees. In fact, they were almost told not to, so it was a very sensitive situation. My intuition, my gut feeling, was that I would be working with them once again. So I think that when I talked to them, they knew what I wanted to do, but it was just a matter of what they could do. It's really funny because I tell my husband that I can just look at Tom and know what he needs before he even tells me. I mean, it's that kind of communication level. And I tease him. I say, "I've known you as long as I've known my husband. I know you well enough to know what you're thinking and what you need." And a lot of times, what he appreciates is that I anticipate what he needs, but only from the years of working with him. I can kind of bring him things that I know he is going to need before he even knows he needs them. It just cracks me up. I know Tom's schedule better than anybody. This is really dramatic, but he's got twelve hours of meetings in an eight-hour day, and yet he finds a way to just kind of smoothly run in and out of these meetings and get done what he needs to have done. But then, in addition to that, he has to account for the interruptions throughout the day: people need this, need that—right away!—as soon as possible!—teleconference! And he finds a way to make that work. I guess you could honestly say, I find a way to make it work because it's my job, too: "Tom, you've got three minutes here, and I think you can do it in between. But you really need to eat lunch. Can you eat lunch in your office while you're taking the conference call?" And it all seems to work out that way. My biggest challenge is managing his time, but he told me he depends on me to do that and somehow, between the two of us, it works out all the time.

PR: Would you say there's a fluid line between the kind of family relationship and the kind of corporate relationship? One of the things that's come up in what you just said, and what other people have said, is that Tom pays an almost parental attention to people.

GD: Oh, definitely! You hit the nail on the head. Exactly. I think Tom is the godfather, and actually I think that's even been a joke here. The godfather of these families; he looks over them. But John is the same way. I see it. John has this huge group. I talk to the assistants throughout the day, every day, and by the end of the day we're surprised: "Whew!

How did he do all of that? How did he get all of that done?" But he makes it happen somehow.

PR: But there's an attention to personal relations that others didn't have.

GD: Oh, yeah. I truly believe that one key to the success that Kathy, Tom, and John have had is in the way they work with people closely. They could be as busy as hell, but it's like, what you have to say is important to them and they always make time. It just cracks me up because I'll see people lined up outside Tom's door, asking, "You got a minute?" And he says, "Oh, sure I do!" And I say, "Are you kidding? Do you see your calendar?" But then I see the appreciation from his staff. That's why Tom is so well informed about what's going on. He makes the time for the information. I have this joke: I say, "Tom, you know the reason why I talk so fast? Because I know I only get thirty seconds of the day to talk to you, and I'm going to fill you with information in those thirty seconds." But thankfully, I had nine siblings in my family, and I know to talk fast if I want to get anything in. So he kind of knows that's the way we both work together. At Roche, when I came in, I felt that this was a family of science. And the way I saw it, the way it worked, the success of it, I think was a result of the way Tom acts as mentor, to young scientists especially. I don't even know if he knows this but I have young scientists that come in to speak with me. They say, "He is mentoring me to do another project, to move into another department." And they go on, "You know what? I know I can do it, because of Tom's guidance. I can't fail." I wonder if Tom even knows this.

PR: He does, and Kathy knows it. Kathy has talked about this a little bit, because one of the things that is very characteristic of her style of management is that she believes that if you put people in jobs that they're not actually really prepared for but they can do, then it is an absolute vote of confidence.

GD: That's what I was going to say. They have confidence that these groups of people will meet the challenge, which makes them more likely to do it. When someone tells me I can't, I'm definitely going to try to do it, but when you've got a vote of confidence behind you, it's suddenly hard to fail. That's exactly how I feel. When you've got them support-ing you, they'll work with you to make it successful.

PR: The term "stewardship" comes from the Greek word for economy, meaning domestic economy, managing a household. For the Greeks, managing a household was not the same thing as managing a business. It entails taking care of the people that are in your domestic sphere as family. And that's a corporate management style that Tom, John, and Kathy practice and seems to work very effectively.

GD: It's a fact.

PR: Is there a downside to it? People in the academic world are very critical of the corporate world. I'm trying to be careful not to be a spokesman for this company.

GD: Yeah, I think there is a downside, because I feel like everyone is my friend here, but I've learned that this is a business. There's a fine line, because I have to be very conscientious in my work and be attentive to the confidentiality of the information; I think I'm very cautious about sensitive information. But I wonder where people are coming from when they ask me questions. Sometimes people have a tendency to want to get friendly, and I have to be sensitive to that and ask myself, "Okay, where is this going? What do they want?" I am very careful about whom to really trust with what information, and anything confidential I don't disclose. Maybe I was raised that way, but it's also a work ethic that I've developed over the years working with management. There's a fence, and you can't cross over that fence. But what's interesting with Tom is that he draws that line, but he also bends back and forth a little bit either way, because that's his way of building trust with the employees that he works with, too. It seems to only be a benefit, as far as productivity is concerned. But even when it's Tom and me, it always has to remain professional. I love that guy like family. You know, I kind of get choked up thinking about it because he's just so endearing. I mean, I've grown up in a big family, a big, large Hispanic family with friends and stuff. Tom? He took a chunk of my heart working for him.

PR: He's the ultimate WASP, too, the quiet Protestant. Understated, careful, attentive, driven.

GD: Exactly! And I think that is what I love about him! He's not arrogant. To me, in my mind, and I tell people this, Tom's amazing! I mean, and it's not just science, it's the way he's figured out people, the way he's figured out how to progress in critical situations. I tell my husband, "Tom is the next Einstein." And I did tell him just a few weeks ago that I've seen how he's come around for people even when they need him in a personal sense—he'll be there. I'm talking about colleagues here at work, even when there's been maybe some personal crisis. He's been supportive, but he stays on track with professionalism and with work.

PR: Yes, he is faced with the twin demands of keeping the whole economy going but also being attentive to people and personal concerns at the same time.

GD: Yeah. But that's like a gift, because I've worked in this field for many years, and you don't see people like that. I remember at Chiron you had to go through ten people to get a decision made. It just wasted time: it was so micromanaged it made me a little bit nervous—it just did.

PR: Okay. What I'd like to know now is how things are working here with this very high-risk, exciting enterprise. How are you experiencing it as well? You didn't step into an ultrasecure job.

GD: It was interesting, because I thought about that. I said, "Okay, Tom, number one question here is security in the company. How do you feel about this?" But he told me about all the challenges that we are about to face, and somehow, even as nerve-wracking as it could be, I thought, "That's exciting!" When he's telling you this, it just sounds challenging and exciting, but the big difference was that we had support from the other companies. That's why I felt even more comfortable than I was at Roche at that time, where I had job security. I would probably still be working there, but the thing is that it was the ideas of the way the company is run. Do I just stay because it's a steady income? Okay, I'll just come in like a robot, work, and go home. That was my thinking about my job. I thought, "No, it can't be that way. I'm going to take the risk, because Tom and Kathy don't want this to fail, and if I can be a contributing factor, I feel better about it; I do."

PR: Do you have a sense that Celera Diagnostics is over the hump yet? Or a few more months? Or a year?

GD: We're on the hump. Yeah, that's how I truly feel it is. I see them work like dogs here. I mean, we've got deadlines, and Kathy's adamant about those deadlines. It's not like we have some flexibility, and the teams here know we don't. It's a fast-working field. That can't be emphasized enough. We just have to be faster. We have to know what we're doing. We have to work quickly, and they've got that in our brains. It's like a rat race: who's going to get to it first? And we have to do whatever we can, and they seem to be providing us with everything we need to make it work, because they know it's high-risk, and we have to get some positive results.

PR: Do you think people get a sense that if it does succeed that everyone will also profit from it?

GD: I don't think so much about that because this whole industry is risky. Five or ten years ago, everyone was like, "Whoa! I want to be rich!" In this economy right now, nah! [*laughs*] I don't really see that. That's not, like, my reward at the end of this trail. It's a nice thought, but I guess I'm thinking more of Celera making a statement, making a mark, showing the industry what we can do and what we are. I mean, that's the first step, and then, as we continue to do that, then we'll reap the rewards.

Interview with Kathy Ordoñez, February 14, 2003

Kathy Ordoñez is soft-spoken, uniformly polite, yet direct in her answers. One gets the sense that she says exactly what she wants to say. The tone of the interview was both orchestrated and cordial.

I asked about her relations with Craig Venter; she responded with a series of positive adjectives. I summarize this portion of the interview. Craig was: "Excellent. Gracious. Supportive. Responsive. Terrific. Scientifically helpful." He was restless and wanted to move on to the proteome and explore variations in proteins. It was no secret that there had been a conflict with Tony White over commercial strategies. "I prefer not to be in the media. I want to do things that make a difference. If people recognize it, that is nice. I was proud to be a colleague of his." Tony White was aggressively looking for someone with pharmaceutical experience, as the corporate decision had been made to go that way. He called out of the blue and that was surprising because she did not have that kind of experience. They had a meeting in New York. He was "Straightforward and direct. Practical and analytic and organized and made things happen quickly." Although Celera Genomics had been built around Venter, Tony White thought that she could pull a new organization together. "I am good at creating an environment where really smart people like to work. I can recognize big ideas and push people to get over the hurdles. It was time for a change in my life, a new challenge." She accepted the offer.

The Eureka Moment

KO: When we came to Applera, the sequencing of the human genome was about to be completed. Of course, there would be a profound impact ultimately on diagnostics, and medicine as a whole, once associations could be made from the genetic information. Other people who watched us in the industry probably think we crept out of Roche with some idea in our pockets, and that is not the case. We just knew we needed to be in a different environment, where we could think more freely, and it was amazing what we went through in that transition. The first couple of weeks were almost terrifying, because we didn't have the pressures of supervising hundreds of people and dealing with day-to-day crises related to products on the market or manufacturing issues or issues in R&D. We suddenly just had time to think! It was very exciting and somewhat intimidating, because if you are accustomed to the type of jobs we had had for years—you came in each morning and there was one hundred and fifty times more to do than you could realistically achieve in one day. So, suddenly we had a blank slate. So we started reading and thinking. And we didn't even have offices; Applied Biosystems gave us space in Foster City, but that was very inconvenient for us to reach. Sometimes we met in my living room, or sometimes we went on walks, or whatever, so we could begin to share information.

One of the things we agreed on was to come up to speed on the detection and other technologies that existed at AB. Tom and John took that on and thought about it. I began thinking about how we would work in molecular diagnostics with a focus on genetics—how could we build a business strategy? We thought there were really two different ways we could go. One would be to just jump into the next generation of testing—that would be genetic testing. Or, we could try to participate in the existing molecular diagnostics market that was dominated by Roche and was primarily focused on infectious diseases and use that as a way to leap into genetics. And there were advantages and disadvantages both ways. The new genetic tests we knew would ultimately take off and be successful, but the uptake curve for them would probably be slower than what we thought we could achieve by taking a piece of an existing market from Roche or Abbott. So we decided that we would try to get into the existing market and use that time, while we were generating revenue in existing markets, to build a very large-scale discovery in genetics. This was important because it would impact the scale of the company and how much research there would be versus development. Our goal, which I had agreed upon with Tony White, was to try and break even in the fiscal 2005–6 time frame.

So with the basic business strategy coming together, we could see the size and scope and approximate how much money we could invest in the research effort. We came up with a pretty sizable amount. As long we were confident of having sufficient success with the existing Applied BioSystems tests that were put into the joint venture, plus being able to take a slice of the existing molecular diagnostics market, that money was secure. And with the scale in mind of approximately how many people we could hire and how much money we could spend, it was all coming together at about the same time.

Tom and I heard Craig give an internal talk about the human genome sequence in early February of 2001. I remember we were sitting in a big meeting at a table that was shaped like a *U*: I was over here, and Tom was all the way over there, and Craig was showing slides that he used when he talked publicly about the human genome. One of the key points that he made—that was just mind-boggling—was that there were just 26,000 genes. Up to that point we had always thought that there were 100,000 or 150,000 genes. I remember thinking, "Gosh, the magnitude of what you really have to interrogate on a genome-wide basis is really significantly smaller than what we had thought." I filed that thought away in my mind to think about later. Several of the key points Craig made about homology and genes being reproduced around the genome were very interesting and were not what I had expected.

I didn't get to talk to Tom that day, or I think even the next day, but we were on the company plane coming back, and Tom and I started talking about how amazing and unanticipated some of the things Craig had said were. That is how it evolved. Imagine: with only 30,000 genes, you could interrogate them with 30,000 experiments, and if you want to look at several different places, you multiply by four or six.

People who are close and work together over a long time sometimes
don't need to talk in sentences anymore. It's like that with Tom. It's
electric: "So what about this?" and "What about that?" and "Gene
chips!" and "We will do this and do that." I just remember looking up
at him and saying, "Aha! We could!" and he looks at me. We have
worked together for so long, but the way we think and process information is very different: I am a very intuitive person, and somewhat
mathematical and analytic, and so it was absolutely apparent to me. If
the answer is there, I see it first and then I go back and derive it. You
could just see Tom's mind going; he was calculating, calculating, calculating, and he said, "You're right." So the two of us got so excited, and
he called John and gave him the same sort of set of data that brought
us to the conclusion that we came to, and John just immediately said
the same thing.

We vetted the idea with an external consultant and with Mike
Hunkapiller. We met with Craig, and we talked with management.
Craig is a very brilliant scientist. He bought the idea of addressing the
problem on a genome-wide basis. We had been thinking of this in
terms of fishhooks and fishnets and how could we capture more and
more. I went home that night—a Friday night, I think—and I could not
sleep. I like to get up very early anyway. I sat at my computer and wrote
a message to Craig to explain to him what we had thought up: Forget
the fishhooks and the fishnets; we are just going to drain the lake and
walk out there and pick up the fish.

We explained it to Tony White and Mike Hunkapiller and the CFO
of the company, and they said, "Go. How much will it cost? How long
will it take? What are the odds that it will be successful?" They had
confidence in us. There was trust.

The second thing that happened was that as we were building our
assays and planning the methodology that we would use for these association studies, I felt that there was something that was missing. I
couldn't exactly put my hands on it. One day I was in Foster City talking
with Mike about the need to have a better understanding of polymorphism. I remember driving back from Foster City and feeling really agitated that there was something missing. We were just beginning
to hire people so it was the April–May time frame, and I remember
going to the blackboard and telling John there is something missing

here; tell me what it is. Resequencing. Mike was thinking the same thing. John went at it, and then Tom came in and they went back and forth on the resequencing effort that I believe will become extremely important. It will show how important it is to have identified so many novel functional SNPs that we can use now in our association studies. Association studies alone are not enough. Finding "the gene for schizophrenia" or whatever is not enough. We think we have a huge advantage. We will do our association studies with thousands of cases and controls, and we will replicate them with two or three or four sets of samples, so that we are not just looking at ethnic differences or whatever. But beyond that our whole plan was to take the information we had developed and configure it into the diagnostic product. I spent my whole career doing that, bringing products to market, figuring out how to create demand for them, educating physicians to order the tests, get the tests up and running in the laboratories. So there was a very coherent business strategy supporting the strategy for discovery.

PR: What enabled you to see this?

KO: The approach we are taking is not in and of itself that important—the scientific approach—but it is the timing of it and the scale at which we are doing it and the environment we are working in that allows us to drive to the next thing. It is the combination of the things we are doing that counts. Certain aspects of what we are doing have been published before, certain aspects of sample extraction, et cetera. So you could go all the way through and say I have seen something very close to this and very close to that—but it is the way in which it comes together and gets aggregated so that we can look at genotyping, expression, and ribotyping and do that on a scale that, to our knowledge, no one else can imagine!

Chapter Two

THE STATE OF THINGS AT CELERA DIAGNOSTICS, EXPLAINED TO INVESTORS AND TO ANTHRO- POLOGISTS

A company can communicate with investors through any of several different genres. Each genre has constraints and preconditions that substantially define its boundaries, and each requires a particular set of skills for its successful performance. One constraint, however, is common to all genres of formal communication with potential investors: a rather impressive series of disclaimers about the truth content of what is presented.

The boilerplate is provided by the Securities and Exchange Commission. Thus, much of this cautionary language is basically identical from company to company. These disclaimers operate in a discursive space where the subjunctive reigns: things may or may not be what they seem;

things may or may not turn out as proposed; there may well be unexpected events or unexpected contingencies. Everything we say, the boilerplate language insists, should be taken with the utmost caution; it is promissory and expectant, not deliverable and existent. The language is so formulaic that experienced investors are unlikely to be driven away from any particular offering by these extended provisos.

For an outsider, these disclaimers are amusing to read because they are much more systematically skeptical than the most severe critic of the industry. The press release that accompanied the December 2002 investor meeting convened by Celera Diagnostics contains a section that reads as follows:

> Certain statements in this press release are forward-looking. These may be identified by the use of forward-looking words or phrases such as "believe," "expect," "intend," "anticipate," "should," "plan," and "potential," among others. These forward-looking statements are based on Applera Corporation's current expectations. The Private Securities Litigation Reform Act of 1995 provides a "safe harbor" for such forward-looking statements. In order to comply with the terms of the safe harbor, Applera Corporation notes that a variety of factors could cause actual results and experience to differ materially from the anticipated results or other expectations expressed in such forward-looking statements. These factors include but are not limited to (1) Celera Genomics expects operating losses for the foreseeable future; (2) Celera Genomics' reliance on Applied Biosystems' emerging Knowledge Business for incremental revenues to Celera Genomics from the Celera Discovery System and Celera Genomics' related information assets; (3) Celera Genomics' and Celera Diagnostics' unproven ability to discover, develop, or commercialize proprietary therapeutic or diagnostic products; (4) the risk that clinical trials of products that Celera Genomics or Celera Diagnostics do discover and develop will not proceed as anticipated or may not be successful, or that such products will not receive required regulatory clearances or approvals; (5) the uncertainty that Celera Genomics' or Celera Diagnostics' products will be accepted and adopted by the market, including the risk that these products will not be competitive with products offered by other companies, or that users will not be

entitled to receive adequate reimbursement for these products from third party payors such as private insurance companies and government insurance plans; (6) reliance on existing and future collaborations, including, in the case of Celera Diagnostics, its strategic alliance with Abbott Laboratories, which may not be successful; (7) Celera Genomics' and Celera Diagnostics' reliance on access to biological materials and related clinical and other information, which may be in limited supply or access to which may be limited; (8) intense competition in the industries in which Celera Genomics and Celera Diagnostics operate; (9) potential product liability or other claims against Celera Genomics or Celera Diagnostics as a result of the testing or use of their products; (10) Celera Genomics' reliance on scientific and management personnel having the training and technical backgrounds necessary for Celera Genomics' business; (11) potential liabilities of Celera Genomics related to use of hazardous materials; (12) uncertainty of the availability to Celera Genomics and Celera Diagnostics of intellectual property protection, limitations on their ability to protect trade secrets, and the risk to them of infringement claims; (13) Celera Genomics' dependence on the operation of computer hardware, software, and Internet applications and related technology; (14) legal, ethical, and social issues which could affect demand for Celera Genomics' or Celera Diagnostics' products; (15) risks associated with future acquisitions by Celera Genomics, including that they may be unsuccessful; (16) uncertainty of the outcome of existing stockholder litigation; (17) Celera Diagnostics' limited commercial manufacturing experience and capabilities and its reliance on a single manufacturing facility; (18) Celera Diagnostics' reliance on a single supplier or a limited number of suppliers for key components of certain of its products; (19) the risk of electricity shortages and earthquakes, which could interrupt Celera Diagnostics' operations; and (20) other factors that might be described from time to time in Applera Corporation's filings with the Securities and Exchange Commission. All information in this press release is as of the date of the release, and Applera does not undertake any duty to update this information, including any forward-looking statements, unless required by law.

Another and different level of constraint to the genre—and this is where rhetorical skill is displayed—lies in the presenter's ability to suggest what the company is doing in enough detail (scientific, technological, patent, and fiscal) to be persuasive but not so much that competitors would profit from the information. Potential investors, old and new, must be seduced, as it were, but, as is in all artful seduction, only so much is revealed. Thus, we are not dealing with dishonesty but rather with a coded performance. Such performances require skilled interpreters or critics. The audience must be attentive and cautious. It must also learn enough to be able to pass along a suitably modified version of this rhetoric to other investors, large and small, not present at the original performance. Finally, the audience may be composed of experienced investors who are not scientifically well-informed. The scientific understanding that participants possess apparently varies greatly.

In sum, rather than deploy the cover-all term "hype," we would be better advised to take these investor presentations on their own terms. In what follows, we summarize the Celera presentations in a form that stays close to the style of the event and its distinctive language. Although there was a live audience, the presentations were also available in real time on the web. Thus there was a virtual audience as well.

Presentation to the Investors, December 17, 2002

On December 17, 2002, Celera executives presented their overview of the company's progress at an investors' day meeting. The events began with the expected legal disclaimer regarding "forward-looking statements": the company is not responsible for updating the information presented, nor is it bound by law to live up to its projections.

The first speaker was Applera's CEO, Tony White, who gave a brief progress report before introducing Kathy Ordoñez. Tony White summarized the company's achievements in the preceding year. In April 2002, Celera had licensed genomic databases and related bioinformatics products, allowing the company to concentrate fully on discovery and development of pharmaceutical products rather than the difficult and time-consuming task of information management. In the same month, by appointing Kathy Ordoñez, who was already president of Celera Diagnostics at the time, to head up

Celera Genomics as well, Celera consolidated a major management role in the genomic and diagnostic companies and moved toward tighter interaction between its genomic and diagnostic facilities and projects. According to Tony White, Ordoñez had already assembled a team of experts in business and pharmaceutical development at Celera Genomics that was instilling "a culture of cooperation" in the reorganization of the company after Venter's departure. The goal was to maximize the potential integration of the most advanced techniques in three key areas of interest to Celera: research on the genomes of different species, the study of protein products coded for in genes, and the management of biological knowledge using computer technology. These three fields are what presenters refer to respectively as genomics, proteomics, and bioinformatics. The interaction between these diverse fields requires compatibility between Celera Diagnostics and Celera Genomics, so that knowledge, research, and technology at both companies can be interwoven in a combined discovery effort. As far as the company's future was concerned, Tony White explained that it was the partnership between Celera Genomics and Celera Diagnostics, and the coupling of genomics and diagnostics in the course of research, discovery, and development, that represented Celera's most promising strategy, adding that this strategic coupling was aimed at achieving something called "targeted medicine." (Ordoñez's later remarks explicate this aim in more detail.) Tony White explained that the idea of targeted medicine was not new but that Celera, with its combined capabilities in genetics and diagnostics, was particularly well situated to make targeted medicine a reality. He also acknowledged that the achievement of targeted medicine represented an ambitious long-term goal, and that more immediate assessment of the company and its stock price would be based on a number of smaller projects.

Next, Tony White gave a quick update of Celera's progress in areas which had been discussed in a similar meeting in December 2001. Celera Diagnostics was on its way to completing three disease association studies and had gained access to patient samples and clinical data associated with Alzheimer's disease, breast cancer, and cardiovascular disease. Additionally, the company had just received FDA clearance on a sequence-based test for mutations in the HIV virus.

Kathy Ordoñez, president of Celera Genomics and Celera Diagnostics, spoke next. Ordoñez explained the idea of targeted medicine more thoroughly, as well as its strategic advantages: The management of Celera

Genomics, in developing a business strategy for a changing company, realized that Celera Diagnostics' disease association studies could be used for therapeutic research at Celera Genomics as well. The goal would be to pinpoint root causes of disease and understand the processes that lead to their development. Celera would also be researching subclassifications of complex diseases that are thought to have multiple genetic causes occurring at different sites in the DNA. Learning to differentiate between different genetic causes for like symptoms would be the first step in uncovering more exact targets for novel diagnostic and therapeutic methods. If targeted medicine succeeded, the result would be a reduction in incorrect diagnoses, hit-or-miss administration of drugs, and adverse responses to treatment, as medicine would be able to address specific causal genes.

Ordoñez emphasized Celera's strengths in the pursuit of targeted medicine: Through the Applera Genome Initiative, Celera had gained access to the discovery and documentation of thousands of novel single-nucleotide polymorphisms (SNPs). (SNPs are variations among individuals in single DNA base pairs located at specific positions in the genome. These variations may be associated with disease.) In terms of technology, Celera Diagnostics provided high-throughput genotyping and gene expression capabilities. Finally, unlike the large biopharmaceutical companies, for whom targeted medicine had been an afterthought forced onto existing research facilities and management teams, at Celera this approach was foundational and constituted the logic of its administrative and scientific organization.

Next up was John Sninsky, vice president of research for Celera Diagnostics, who began by explaining that the human genome project was the beginning, not the end, of research into the potentiality of genomics in medicine. The initial research into the genome had changed the starting point for biopharmaceutical companies: the human genome was now thought to have approximately 30,000 genes, revising original projections which had placed the number at around 100,000; many genes could be expressed in several ways, meaning that the same gene sequence could produce multiple proteins; almost half of predicted genes had no known function; genes were found to be unevenly distributed along the genome. Disease complexity, then, could be attributed to both variations in the genetic sequence and to the intricacy of gene expression.

These changes in the understanding of the genome shaped much of the thinking and subsequent strategizing at Celera Diagnostics and convinced the company's executives to move away from older approaches. Sninsky moved rapidly through a list of problems with several approaches to disease association studies (each of which represented the strategy of a competitor). According to Sninsky's brief comments on three of these approaches—linkage, linkage disequilibrium, and haplotyping—the core difficulty lay in pinpointing actual genes. These techniques, although they generated vast quantities of information about the segments of chromosomes that contained disease-causing SNPs, were proving ineffective in locating disease-causing SNPs within those segments. Celera's approach, Sninsky said, would be different. Disease-causing mutations altered, in many cases, the coding capacity or the regulatory and splicing regions of genes. Key mutations affected which proteins were being produced, at what levels of activity or stability, and in what amount. Celera's strategy would focus on SNPs believed to be functional in altering proteins, starting with known portions of the genome and moving toward an entire genome scan. In a sense, Celera's approach would attempt to get straight to the root of disease-causing genetic mutations.

The Applera Genome Initiative, a project involving researching and cataloguing information about the genome, represented a key preparatory step for a scan of functional SNPs. The initiative consisted of three major steps: it sought to resequence the genes and regulatory regions of thirty-nine individuals and one chimpanzee; confirm the mRNA transcripts of predicted genes to establish which proteins were actually being expressed and produced; and then identify the medically relevant mutations. But why sequence thirty-nine individuals and a chimp? Sninsky explained that it is currently unknown whether rare mutations (those that occur in 1–5 percent of individuals) or more common mutations (those that occur in 5–20 percent of individuals) play a role in causing complex diseases. Sequencing thirty-nine individuals would facilitate the detection of both rare and common genetic variants. Sequencing the DNA of a chimp might uncover evolutionary links between primate species that would point to disease-causing alleles.

Sninsky went on to say that the Applera Genome Initiative had already made significant progress. The resequencing of coding and regulatory regions in 23,400 genes had been completed. Coding SNPs had been identified and catalogued. Similar work on regulatory SNPs was expected to be completed in January 2003. Lastly, the discovery of mRNA variants was projected to be completed in June of the same year. In sum, Applera was close to isolating the functional mutations that cause amino acid substitutions and therefore affect the production of proteins. This identification of functional SNPs would then be used to both stratify patients according to predispositions to diseases and drug reactions and monitor patients by profiling gene expression over time.

In the second half of the meeting, Tom White, chief scientific officer and vice president for research and development, addressed the potential role that Celera's genomic tools could play in bringing about targeted medicine, emphasizing that, where targeted medicine has succeeded, it has become the standard of care. He used examples to highlight how targeted medicine had already been put to work in Celera Diagnostics' own HIV diagnostic test, which identifies mutations in the HIV virus that cause it to be resistant to certain drugs. With the ability to identify causative mutations in HIV comes a new understanding of case-specific therapeutic approaches, tailored to the genetic peculiarity of a specific strain of HIV.

Tom White went on to describe some of the studies already underway at the time of the meeting or scheduled to begin soon thereafter. His theme was efficiency: Celera Diagnostics, because of its technological advantages and the scale on which it operates, is able to move through the multiple steps of a study rapidly. For some diseases, such as Alzheimer's and arthritis, linkage regions are already well characterized, so that all the genes in those regions can be studied. Additionally, by considering both genotypes and gene expression (which also entails an environmental component), Celera is able to recognize drug efficacy and toxicity earlier in the course of clinical trials than ever before.

Next, Tom White discussed host-response studies: Celera Diagnostics' research focus is on genetic markers that affect a patient's response to an entire class of drugs. As more knowledge becomes available about the specific genetic mutations linked to manifestations of disease in different patients, diagnostic tests for these markers would likely alter therapeutic choices. Additionally, identifying markers that do not respond to existent classes of drugs allows researchers to develop new drugs specifically for those patients who are resistant to current therapeutic options.

White gave two examples of the potential applications of targeted medicine: Patients infected with the hepatitis C virus have very different rates of disease progression, causing variability in the pace at which the condition of the liver worsens in untreated patients. Pinpointing the genetic cause of this variability, whether genotype, differential gene expression, or both, would constitute a crucial first step in the development of a much-needed molecular diagnostic tool. The kind of test that emerges from this line of research would provide some criteria for distinguishing between cases that require urgent, and sometimes difficult, treatment and cases that can be addressed with a milder form of therapy.

Furthermore, expression markers or SNPs in a panel of interferon response genes might be used to predict how a particular patient will respond to interferon, the most common hepatitis C drug therapy. Entering the market for diagnostic tools that provide host-specific information seems all the more appealing when one considers the price of drug therapy, particularly for hepatitis C. Identifying a patient unlikely to respond to the drug means saving her the cost and extreme stress of undergoing ineffective treatment, as well as the discomfort of possible side effects.

The two diagnostic tests—one for the rate of disease progression and the other for drug efficacy—can then be used to position a patient along a graph defined by these two axes and determine appropriate treatment. For example, patients who have a disease that progresses slowly, whether they are drug responders or nonresponders, could safely delay treatment or avoid certain treatments altogether; in contrast, patients with fast-progressing liver damage who are responsive to the drug would have good reason to undergo harsher treatment. Finally, those patients with fast-progressing hepatitis C who do not respond to current treatments could be isolated as candidates for experimental drugs that are currently being tested in clinical trials. This last group of patients would be benefited by the opportunity to try new classes of drugs as they develop, while their participation in studies would allow researchers to identify specific targets for drug development.

With the added capacity to classify patients according to differential genetic risk of adverse response, Celera Genomics would be able to test drugs on high-risk patients in the first stage of clinical trials and thus avoid costly and time-consuming surprises later on.

Presentation to the Anthropologists; January 3, 2003

On January 3, John Sninsky met with us to explain the four approaches to disease association that had been touched upon at the investor meeting. Excerpts from the January interview appear below. Presumably because of the very different audience, Sninsky's presentation on this occasion contrasted markedly with the one he gave at the investors' day events and demonstrated his remarkable versatility of style. Equipped with a white board rather than a set of prearranged slides, Sninsky proceeded to give us what can best be described as a magisterial lesson on the material at hand. He graciously responded to our questions, adroitly facilitated an interactive process, and consistently made sure that we were still following him throughout the two-and-a-half-hour tutorial. When he named specific companies in association with methods being criticized, he did so only upon Rabinow's request and with evident hesitation. It seemed that Sninsky felt more comfortable criticizing an approach as such, and the occasional outspoken individual proponent of it, than the companies or institutions behind it.

Linkage
Sninsky began by explaining that population size and number of generations are major strategic considerations when choosing subjects for research into disease-causing genes. Bear in mind that what disease association studies aim to do is to identify specific mutations that are causally related to the predisposition for, or actual incidence of, specific diseases. In order to find these mutations, researchers have to locate them within the genes of afflicted individuals. A small population of recent origin will share large regions of DNA in common, allowing researchers to zero in faster on the particular region where the disease-causing mutation is located. Such an approach, referred to as linkage, is being pursued at deCode Genomics, one of Celera Diagnostics' major competitors. In this case, genetic information is taken from the Icelandic population, which consists of approximately 270,000 individuals. Both the size and founding date (around A.D. 900) of this population provide a good starting point for linkage research. Furthermore, Iceland's population experienced two catastrophic events, a volcanic eruption and a famine (in the eighteenth century), which reduced the amount of variability in the regions of DNA represented within the population.

A major point of divergence in disease association studies lies in the number of genetic markers used. Short tandem repeats are sequences of DNA that repeat themselves, for unknown reasons, throughout the genome. By highlighting these regions, scientists are able to gauge the relative location of DNA sequences within genes and chromosomes. Currently, each marker costs approximately one dollar to locate. Hence, using many markers in studies that have to be replicated for verification purposes is expensive. In this respect, deCode's method is cost efficient: linkage uses only about four-hundred markers across the entire genome to locate regions containing disease-causing SNPs. Sninsky explained that markers are like grates on a filter. If the grate is too wide, that is, if there are too few markers, the regions in which mutations inhere fall through. This means that in a large population, many generations removed from its origin (such as the population of Europe), four-hundred markers would not suffice for the purpose of finding the relatively small regions of shared DNA.

On the other hand, dealing with entire regions of genetic material has a definite downside:

JS: The difficulty deCode is having right now is finding the gene that causes disease. And the reason that's difficult is: (1) because of the number of genes that are possibly causing disease, and (2) that, on average, there is about one genetic change every thousand base pairs. So there are a large number of genetic alterations in there, all of which could play a role in causing disease. So Kari Stefánsson [founder of deCode] has the advantage that the portions of the genome are larger, so he can use a small number of markers, but he has a disadvantage in that he has a hard time finding the actual genes when he gets the regions.

PR: Before the genome mapping there really was no way to decide on the best method. The tools weren't there, the sequencing power wasn't available. So in traditional genetics, each of these approaches made sense.

JS: That's an excellent point, Paul, because before we knew the genome, linkage was clearly the best way to go.

However, in the context of a dynamic process of discovery and innovation, where the baseline knowledge changes rapidly, a particular approach can seem promising at one point and much less so soon thereafter. Models and definitions shift constantly. Sninsky rehearsed some of the changes in the common conception of the genome that had occurred over a short period of time: genes are now thought to be discontinuous; the same coding region can generate multiple proteins; some diseases that were formerly characterized according to their symptomatic expression are now being found to represent multiple diseases caused by different genetic mutations. In constant motion, genomics research exemplifies the tension between incremental change, which incorporates new notions into existing research models, and a more radical abandonment of "the old way of doing things" as large obstacles arise, in favor of entirely new approaches. DeCode has been practicing the former, meddling with linkage in order to get results. For example, scientists at deCode are adding a filter for causative genes that will concentrate on gene expression in specific parts of the body: for a disease that manifests itself in the brain, such as Alzheimer's, the researchers look for genes that are expressed in the brain. According to Sninsky, though, given the limited success with existing models, the latter approach, while potentially more fruitful, is undoubtedly hard.

Linkage Disequilibrium

Linkage disequilibrium (LD) uses many more markers than does linkage. The logic behind this approach runs as follows: If the number of markers is increased, smaller segments of shared chromosomal DNA can be located. To return to the filter analogy, if you decrease the distance between the grates on the filter, you can isolate smaller segments. Linkage disequilibrium, therefore, enables researchers to study large populations that were not founded nearly as recently as the population of Iceland. Furthermore, locating smaller segments of chromosomes restricts the number of mutations in each segment, thereby decreasing the number of possible causal mutations. Debates on the right way to approach linkage disequilibrium abound.

JS: A long and involved discussion that's still outstanding is, How narrow do the grates need to be? And most people are pretty happy with a number of 300,000 to 500,000 markers, instead of 400; and that presents a technical challenge, because this number is so large. If it costs you about a dollar per marker, and you have to look at 1,000 patients, then all of a sudden you're talking about hundreds of millions of dollars to do a study. The advantage is that, if the genome is 3 billion base pairs long, you're dividing it into 300,000 segments. You're cutting it into still fairly large chunks of DNA, but these segments are much, much, much smaller than the ones that are isolated using linkage.

So there were some people that were proposing this, and there are still some people who are proposing it. But they realized that this number and the technology still don't get them to where they need to be. So some other people said, "Aaah . . . I have an idea."

Haplotyping

The idea was a method called haplotyping, an alternative strategy that involves finding shared blocks of DNA that are common to all humans. Under-
pinning this approach is a growing body of evidence supporting the theory
that the modern human population was actually founded quite recently
in evolutionary terms and was initially small. Using this information, researchers designed a method for finding markers that are predictive of small
blocks of DNA shared by all humans. Unlike linkage disequilibrium, which
places markers throughout the genome, in order to divide chromosomes into
manageable sizes for research in large populations, haplotyping aims to mark
only specific blocks. In addition, since the human population is much older
and larger than the Icelandic population, these blocks are much smaller than
the blocks studied at deCode using the linkage approach. Sninsky explained:

JS: There were two paths out of the problem with linkage disequilibrium,
    that are still being followed. The one path is to make the technology
    cheap enough so that we can do this, and people are running down that
    path. The other path is to say, "Do we really need 300,000?" And the rea-
    soning comes back to those blocks of DNA that we talked about. Because
    it turns out that we humans actually haven't been around, relatively
    speaking, very long, and human populations were founded on a rela-
    tively small number of individuals. Right now it is assumed that the
    human population started with about 10,000 founders, or that there
    were major bottlenecks in the history of the human population, but it
    looks like the DNA that we have was only contributed by 10,000 indi-
    viduals. As far as we can tell, Neanderthals haven't been involved. They
    were a separate species that didn't contribute to our DNA. So, some
    people have argued, we have a relatively small founding population, and
    we're not as old—arguably, we're only 150,000 years old, because that's
    when there was a massive expansion of the population. So, according to
    this logic, maybe there are portions of the chromosomes that are shared
    blocks that we can find. Instead of taking a picket fence approach, we
    could take a block or a haplotype approach. What if we just find mark-
    ers that are predictive of the blocks? Now these blocks are much smaller
    than the blocks from deCode because the number of individuals that
    founded the human population is a lot larger, and the number of gen-
    erations is also a lot larger. So the haplotype approach is being pushed

by multiple companies, but probably the strongest supporter is Eric Lander at MIT.

When we looked at these markers, well, some people thought that we need 300,000 to 500,000 markers for linkage disequilibrium, and then when you read the newspapers, they say that what's interesting about the haplotype approach is that you only need 300,000 markers. So your first inclination is to say, "But wait a minute, that's not a reduction in number. That's the same number." The difficulty here is that as science proceeded with different populations and new knowledge about genetic variability, the number of markers for linkage disequilibrium went up to about 1½ million. So what happened was that they ended up with the same number of expected markers and they claimed it was a success, but in fact it wasn't, because they went through a period where they thought they needed more than a million and a half markers. Technologically speaking, they are still in trouble. They still can't get to where they need to be. So the U.S. government, with the completion of the human genome, has decided to support, to the tune of $100 million plus, what's called the "Hap-Map." I'll give you one qualification. They said, "You don't need to find the mutations; you just need to find the blocks." In the simplest case, when you say the chromosome is made up of blocks, the first thing that comes to mind is that you have these adjacent blocks that make up the entire human genome. Well, it turns out that when you go back and take a look, 150,000 years and those 10,000 individuals actually introduced more variability, and only about half of the genome is probably covered in these blocks. The other half of the genome now is lost. Actually, it's not lost; it's mixed up to the extent that it is no longer a block. So it may be that this haplotype scan is a half a genome scan, because it only looks for the blocks that we know exist.

It's not an accident that the U.S. government has decided to do this, because Mr. Lander is a very high-powered politician, with powerful allies, and was a big player in the mapping game, heading up the Whitehead Institute, which is a privately funded institute that's connected to MIT in Cambridge. The question for the public consortium is, What's the next stage with all this money that they committed to genome mapping? The French are doing something similar, and it's not succeeding very well. And some people have argued that what those supporters of the Hap-Map have done is to proceed as if they have forgotten we know where the genes are. Because what they're basically doing is mapping, right? Because the genome is sequenced, and now they're saying, "Don't worry where the genes are, just worry about where the blocks are." The interesting thing is that the way they got the human genome is by being mappers, and finding chromosomal regions. One can make the argument that these people are comfortable with a mapping paradigm, but they're not ready to step into another paradigm.

There are two potential strategies for managing the cost of markers: First, the price of markers could be reduced if the technology involved were improved. Sninsky noted that though he had heard claims of success in this area, he hadn't seen any proof of a potential reduction in price yet. The second strategy involves pooling patients and controls. For every study done on 1,000 individuals, 500 are patients and 500 are controls. Rather than use every marker in every person, researchers can use the 300,000 markers on two DNAs, one representing pooled DNA from patients and the other representing pooled DNA from the control samples. In such a way, the cost of the experiment can be drastically reduced. The weakness of this strategy, however, is that it increases the difficulty of sorting out causal mutations in diseases that express themselves similarly but can be caused by multiple, independent genetic mutations. The pooled patient sample may constitute a mixture of genetic causes for disease, thus making it hard to separate and characterize which set of mutations belongs to which dysfunctional pathway.

Sninsky compared current thoughts on cost efficiency to similar debates in the sequencing of the human genome:

JS: Some people are saying, "Let's make the technology cheaper." Some people are saying, "Let's live with the existing technology, but let's pool." So both of those approaches have risk involved, because you have to wait for this to occur, and it may not. The risk here is that you are thinking of these pools as one disease, and they probably aren't. Let's come back to an interesting revisiting of history. This very same question was being raised in the human genome. Because the plan for the human genome was, Let's do a little bit of sequencing, but let's find new technology to make it cheap enough per base so that we could sequence the whole genome. And basically what happened is the technology didn't get very much better, and finally people said, "You know what? Let's just do it."

Celera Diagnostics' Approach

The core of Celera Diagnostics' own approach is the gene-centric functional SNP scan. Celera Diagnostics is focusing specifically on SNPs that change amino acids, mRNA splicing, and gene regulation. Although mutations occur all the time, only a few mutations actually change the composition of proteins. Those that don't change proteins are referred to as silent mutations. Of the mutations that do change amino acids, some do not cause functional disturbances in proteins, while others, depending upon their shape and their location, can disrupt protein function. In other words, some mutations are altogether silent, causing no alteration, while other mutations are functionally silent, leaving the function of proteins unchanged. Sninsky described what may be considered the crude version of the science on which Celera Diagnostics' own approach is based:

JS: When we look at the variability between individuals, we actually only find about one genetic alteration, one single nucleotide polymorphism per gene that affects the protein. In brief, a gene produces a messenger RNA that is made up of one of four different bases. We've learned that that messenger RNA encodes or allows for the production of the protein. That protein is made up of twenty building blocks, or amino acids. And it turns out that the way the messenger RNA encodes a protein is that every three nucleotides designate a different amino acid, a different building block. So, there are sixty-four different triplets, because there are four different nucleotides and sixty-four possible combinations of three nucleotides. Some of these tell the protein machinery to stop making the protein. Some of these are redundant. That is to say, multiple different triplets encode for the same building block.

It turns out that for some triplets, for example, commonly this redundancy occurs at the right-most nucleotide. It turns out that ABC, ABD, ABE make amino acid 1. That's a redundancy in the code. So now you can say, well, if there is a genetic alteration between two people, if the genetic alteration occurs in the third position, most of the time it actually won't lead to a change in the protein, because of the redundancy in the code. Those are called silent mutations. The reason they're called silent is because they actually don't appear to have a functional consequence. If, on the other hand, the first two nucleotides were changed, it now no longer would code for amino acid 1; it now would code for amino acid 2, a different building block.

A protein is just a string of amino acids, but it folds up into a ball. Sometimes the amino acid change occurs on the outside of this ball, which interacts with water, and water just accommodates it, so it doesn't change the function of the protein. If that amino acid 2 is substituted in other places in the ball, it can disrupt the protein. So sometimes there's not an amino acid change as a result of a mutation, and sometimes there is an amino acid change, but the mutation is functionally silent. And the reason that it's functionally silent is that it doesn't occur in an important region of the protein. But on the other hand, sometimes mutations disrupt the protein and the ball has to change, and now it is no longer functional. It can't carry out the activity that it needs to carry out.

The eureka moment, as Sninsky recounted it, occurred when Tom and Kathy were flying back from a meeting with Venter in which he had mentioned how few mutations actually lead to alterations in an amino acid. Tom turned to Kathy on the plane, having ruminated over the discussion with Venter for quite a while, and proposed that instead of using 300,000 markers, Celera Diagnostics would use mutations that altered amino acids as their markers. Considering the rate at which these mutations occur, 30,000 markers would be a sufficient number for functional scanning (30,000 genes; one amino-acid–altering mutation per gene). Furthermore, of the approximately 3 billion base pairs in the human genome, differences between individuals (hair color, stature, predisposition to disease, etc.) are accounted for by only about 3 million base pairs. It follows that disease-causing mutations are contained within these 3 million mutations. Sninsky noted that "there may be as few as five hundred mutations that have a functional consequence."

After having come up with a strategy that revolved around functional scanning, Celera decided to sequence the DNA of thirty-nine individuals in order to find and categorize mutations according to their prevalence in the genes of different human beings. Knowing, from the information provided by the human genome project, where genes are located has enabled Celera to limit sequencing to the 1 percent of the nucleic acid that actually codes for proteins. Sninsky pointed out that some of the objections to this approach stem from disagreements regarding the difference between monogenic and polygenic diseases. The former are diseases that originate in one site in the DNA, while the latter are diseases caused by mutations in multiple sites. Celera's scientists reject the view that polygenic diseases are not coded for in the same regions that code for proteins and believe instead that monogenic and polygenic diseases simply represent extremes on a continuum. This means that scientists at Celera Diagnostics expect to use the same detection strategy to uncover both kinds of disease. They have been criticized for this approach by scientists who point out that polygenic diseases, such as rheumatoid arthritis, express themselves only later in the lives of patients, unlike most monogenic diseases. These critics hold that genetic alterations associated with polygenic diseases have less expressivity than those that cause monogenic diseases, thereby illustrating a fundamental difference between these two categories of genetic predisposition. In response, Sninsky maintained that the polygenic diseases express themselves later in life because of genetic redundancy: if there are three genetic pathways through which a particular protein is produced, of which one contains a functional mutation, it isn't until the two normal pathways are worn out (often by environmental factors) that individuals develop the disease.

Of course, whether or not functional scans represent an effective strategy for detecting the cause of any disease will be shown by Celera's ability to produce results. Sninsky seemed optimistic. He also gave the impression that Celera Diagnostics had already gotten encouraging results.

Genes: Old and New

JS: It turns out that the blocks of DNA in the Hap-Map are actually different for African Americans and Caucasians and Asians, so it will probably take them two or three years to complete the map. Celera is doing experiments now. In fact, we got recent results to suggest that we have found an association for a disease called Alzheimer's. We are doing this now, and the first whole genome scan that we are going to do will probably be later this spring, probably April, May. So rather than waiting two years to do what I might call a half-genome scan, we are doing whole genome scans in a couple of months.

Tom White uses this metaphor sometimes for changes in science: once someone tells you where Orion is in the sky, the three characteristic stars of the belt, every time you look at the stars, you could see Orion, but you can't see Orion if people didn't tell you that Orion was there. So what happens sometimes in science, and in other disciplines as well, is once you figure something one way, it's hard actually to step out of that thought process and think of something fundamentally differently. Thomas Kuhn is a very well-known philosopher of science, and he talks about paradigm shifts. And those paradigm shifts are like those three stars of Orion. Anytime you look in the sky, there's Orion, but somebody else doesn't see Orion; he sees those three stars in the context of another constellation. That would be a paradigm shift. And what's interesting is that there are people who, in the context of new information, can't change their paradigm.

So, it used to be that a gene was defined as a discrete inheritable unit. You can argue that, though Mendel did not use these types of terms, that is the way he defined what we now know to be genes: discrete inheritable units. And then we began to have what was called the one gene/one protein paradigm, where one protein was made from one gene. That paradigm led us to believe that the human genome would contain about 100,000 genes. Some people said 150,000, but 100,000 was not bad. What we now know is that there are probably closer to 30,000 genes. But here's the difference: Instead of the paradigm of one gene/one protein, it actually turns out that there is one gene and multiple proteins.

PR: The gene is not a discrete unit?

JS: The gene is not a discrete unit. In fact, instead of it being contiguous, it's non-contiguous. And the reason you can generate multiple proteins is because of the non-contiguity. So that there has been a reduction in the magnitude of genes, the definition of a gene has changed, and the number of proteins has actually increased. What people say now is that there are 30,000 genes; there are 100,000 to 200,000 messenger RNAs, which generate about that same number of proteins; and there are probably a million different proteins because of the modifications of proteins. You can alter these proteins by attaching things to them. You can attach a phosphate to a protein, a carbohydrate, or a sugar.

PR: I notice that every scientist in the world thought that the 100,000 figure was a plausible figure, everybody. What was that false consensus based on?

JS: The number of nucleotides in the genome, which is 3 billion, divided by roughly how large genes were supposed to be. So it could be a bit more or a little bit less, but it was thought that was the ballpark. One of the wonderful things about science is that you learn something. Within

x number of months, or years, there will be an answer to the question, How many genes are there? We could say this is the end of genetics as we know it, in the sense that genetics meant a gene was located somewhere and produced one thing. We now think that both of those are wrong, or at least need to be dramatically rethought.

So, the best way to think about changing the definition of the gene is that it went from a contiguous segment of inherited material to a contiguous segment of nucleic acid to a non-contiguous set or combination of regions of nucleic acid. And now a discrete messenger RNA that makes a unique protein is probably what people think about when they say something is a gene.

(Even More) Technical Exposition: Ribotyping

Kathy Ordoñez strongly emphasized that the distinctive characteristic of Celera Diagnostics' approach is "the way in which it comes together and gets aggregated so that we can look at genotyping, expression, and ribotyping and do that on a scale that, to our knowledge, no one else can imagine." Although the previous sections have been somewhat technical, the concept of the gene and its functions are relatively widely disseminated. The following section deals with the third component Ordoñez mentioned, ribotyping, and this material is even more difficult. Ribotyping is also the most untested and uncertain of the technologies under discussion. Nevertheless, we felt we had to include it precisely because its future uses are unclear and potentially important.

Interview with Shirley Kwok, August 30, 2002

SK: I have a very long-term working relationship with Tom and John. I really respect the work that these two guys have done. I felt that what I was doing at Roche, at the time that I decided to leave, was pretty much dead-end for me. I was heading the Department of Infectious Diseases. It wasn't clear that the company was going to be moving forward with anything in infectious diseases. I know that Tom and John had been pushing for expanding the program, to do host-response to infectious agents. It's a brand new area, it had a lot of potential, and so I decided that I'm ready to take on something new. They were interested in having me start a brand new area. This is a challenge to learn and to help bring something to this organization. I don't really have a strong background in immunology, and so to get talked into this was a little bit overwhelming and still is. I think that Tom and John had a lot of confidence in the people that joined Celera Diagnostics from Roche. They certainly didn't bring us in for the expertise that we demonstrated at Roche in particular areas. Perhaps what they were looking at was our ability to broaden horizons. I wanted a new challenge, and I got that.

We have some different approaches: one is genotyping, one is mRNA expression, and what I'm working on is what's called ribotyping. In ribotyping we are trying to determine the expression level of the two alleles in the host. The fundamental question is, Do differences in expression levels of two alleles provide some indication in disease progression or indicate response to therapy? Do they have anything to do with the pathology of different diseases? So, with that challenge, that in itself has been a technical challenge, my task has been to build that into technology, finding how can I develop an assay that will allow us to differentiate the expression of the two alleles. Up until this point, I've been working on developing the assay strategy. I am not so much looking at the disease, any particular disease at this point, but just having the technology in place so that when we move on to the clinical studies, we can plug in those studies. If this works, then potentially Celera Diagnostics' three approaches work in parallel.

PR: Is anyone else doing this?

SK: No. There have been a lot of publications on genotyping SNPs and genotyping messenger RNA expression. Differential allelic expression is just beginning to appear. I think it's partly because it's so technically challenging that people haven't quite figured out what's the best way to do it. It's challenging from the standpoint that the conditions that are required to make the RNA, to make the cDNA from RNA, the reverse-transcription [RT] step, are not compatible with the conditions that are required for allele-specific PCR. You have two conditions; we want to have this done with a single enzyme that would do both RT and allele-specific PCR, and those two are just not compatible, because the conditions with enzymes that are most favorable for RT are the exact opposite of what's necessary for allele-specific discrimination. We did try to get it to work with a single enzyme, when I realized that we're going to have to evaluate different approaches to get to where we need to be. I think we have a method now that works pretty well. Now we're beginning to do some model studies just to see whether we can detect differences in allelic expression. It's been a year, so I'm not displeased with where we are.

PR: Would this be a general technology that might serve as part of a platform?

SK: Exactly. I don't know whether it would be a breakthrough technology, but it would be another technology that we add to our toolbox, and hopefully together we'll be able to gain some insight. Based on the literature there is evidence that differential expression is playing a role in a large number of diseases. Whether or not that's going to be the case with the specific disease that we're working on, that's what we're trying to find out.

PR: When you say different allelic expressions, does that mean these are mutations?

SK: Yes, generally caused by mutations in the gene—either the promoter or regulatory regions—that cause the allele to be degraded; you have only one allele being expressed, or the amount of expression from it is reduced. The amount of protein ultimately made would be different, and the reduced amount of proteins may contribute to some problems.

PR: So this is a significant step beyond "this is the gene for, or is a marker at the gene." This is actually the site of the functional expression.

SK: Yes, it is a functional site. All I can say is by looking at these two alleles, there is a difference, and that difference can be due to deletions, insertions, mutations, a slew of things. It doesn't tell us mechanistically what is going on.

PR: And how do you choose the locations?

SK: It's really based on the SNPs. We go for high-frequency SNPs.

PR: Are you doing studies to determine what the normal conditions are?

SK: That's what we're doing next. We have to get an idea of what normal is; including variation and what distributions there are.

PR: And if all were to go wonderfully well, where would you be in a year?

SK: Hopefully applying it to some clinical study, which we have picked already. As soon as the samples come in we can look at them. You know we're working together with the genotyping group and the messenger RNA profiling group; I'm tagging on to some of the other collaborations that have already been set up, since we really don't know how applicable this technology will be. I think in six months we should have a fairly good idea as to "go" or "no go." We are going to do the initial studies to see what sort of variation we see. If we don't see enough variation, then we'll have to decide, are we looking at the right population? Do we need to dissect further? Are these differences being mapped? Then the question is: Are we looking in the correct cell populations?

Interview with Shirley Kwok, September 10, 2002

SK: As indicated before, ribotyping is trying to measure the expression from each allele. In order for us to be able to track the expression of each allele, we needed a marker. How can we differentiate allele 1 from allele 2? To do that, we have selected a SNP to track the expressions. What we're looking for are individuals who are heterozygous at that marker. In order to make our studies as relevant as possible, we need to select a SNP that is present at a fairly high frequency. The general method is to find a SNP of a high frequency in a region of interest; the SNP will vary from disease group to disease group. Essentially what we're doing for ribotyping is we're really combining the two technologies of genotyping with message expression. Genotyping from the standpoint that we need the SNP and the SNP will define the final design, and then the RNA expression from the standpoint of looking at the amount that's being expressed. Our goal is to use RT-PCR to determine the relative expression of the allele.

The SNP that we're selecting is not necessarily the causative SNP. This is just a SNP used as a marker, whereas in genotyping groups, they're actually looking for a rare causative SNP that might be associated with disease. If we were to take that approach, ribotyping would be significantly hampered just by numbers alone. When we do the ribotyping, we are measuring the expression of the two alleles relative to each other. Since we're working with heterozygotes, we expect to get signals from both alleles. Allele 1 will serve as a control for allele 2 for example. In mRNA profiling, their expression is compared to a housekeeping gene; it's more of a technical thing. For mRNA profiling, it can be applied to all genes. In ribotyping we are much more restricted by our ability to find a SNP, a high-frequency SNP, whereas in mRNA profiling they can pick any gene they want as long as they can find conserved regions, and we've got the huge genome to work with.

Often we find a case where you may have a mutation in a regulatory region that just lowers the amount of protein expressed—the amount of RNA transcribed and ultimately the amount of protein synthesized. If you looked at this gene by messenger RNA expression alone, you would not see that the expression was skewed to one allele; you would be looking at the total expression of it. This helps us to dissect whether or not the mutations from any one allele could be responsible for what we've seen.

So why are we interested in ribotyping? We think that differences in allelic expression have already been attributed to a number of diseases. And functionally we think that ribotyping is a very powerful tool because it helps us measure the effect of genetic alterations at or upstream of the SNP. If the SNP that we're looking at happens to be the causal, we'll be able to measure what the consequence of that is in terms of the expression levels. But since the SNP only serves as a marker, if there are other events that happen upstream—deletion, insertions, chromosome expansion; doesn't really matter what it is—those differences can affect the expression. And once we find that the expression of the two alleles is different, we can then go back and say that there is clearly something going on.

There may be cases where along the gene there may be multiple mutations, so if you were to go and look at the genotype of this SNP, you'd find that all these, each mutation alone, adds a small amount of risk. But it's possible that all these mutations together are what's important. We might be able to measure the effect of all these mutations, or some of all these mutations, on the expression using this.

Chapter Three

THE
MACHINERY
AND ITS
STEWARDS

In his magisterial book *The Social History of Truth*, the historian of science Steven Shapin informs us that in 1680 Robert Boyle published the second part of his *Continuation of New Experiments Physico-Mechanical: Touching the Spring and Weight of the Air*. This book, Shapin tells us, was in most respects an unremarkable extension of Boyle's researches of the previous twenty years. What *was* remarkable however, was that in the preface Boyle writes that his trusted technician, Denis Papin, had done most of the work. "I gave him the freedom to use his own [pump], because he best knew how to ply it alone, and [. . .] how to repair it most easily."[1] This textual acknowledgment is quite extraordinary, as it was rare indeed for a gentleman to mention the work of an assistant. Furthermore, it is clear from the text that Boyle had given Papin a great deal of latitude in designing the actual experiments themselves: "Some few of the inferences owe themselves more to my assistant than to me. [. . .] I had cause enough to trust his skill and diligence."[2] Boyle assures his readers that he acquainted himself with all of the relevant alterations Papin made during the course of the experimental series and that he supervised the final reports, composed by Papin. Boyle provided his guarantee as to the credibility of the results. There was nothing unusual about a gentleman philosopher guaranteeing the results of work done in his home laboratory nor about Papin's multiple capacities; what was exceptional in the extreme was the fact that Boyle named Papin. In the seventeenth century, natural scientists were gentlemen, with proper names; technicians were anonymous. This system of credit, established during the time of the "scientific revolution," is still basically in place today.

Who were the technicians? Technicians, Shapin writes rather quaintly, are "persons who are remuneratively engaged to deploy their labor or skill at an employer's behest."[3] In the seventeenth century, however, the term "technician" was not used in the context of scientific experiment; one hears, rather, of "operators," "domestics," "assistants," "workmen," "a boy in my employ" (such work was a male preserve), et cetera. Surprisingly, it is only in the twentieth century (and with any regularity only in the second half of the century) that the term "technician" became associated with science. Upon reflection, we realize that as one after the other of the natural sciences came to depend upon experimentation done with ever more complicated machines, each became a practice requiring a division of labor and sizable resources of space, time, money, power, authority. Since we take these conditions for granted, they remain largely invisible or absent in the accounts we provide of scientific research.

Even in the seventeenth century, science was already a collective activity heavily dependent on machines; it was an activity rooted in trusting (as well as overseeing) those who took care of the machines. Within the contemporary human sciences (as opposed to journalistic accounts, in which heroic geniuses make discoveries that everyone else recognizes and hails as true), it is a commonplace that science is a collective activity: of individuals spatially assembled in "a lab"; of larger groups of specialists doing parallel work, who constitute competitors and peers; and of extended networks of diverse sorts, including fiscal, legal, political, and now ethical. It is less well established that precisely because science is a practice dependent on many people doing different, if interrelated, tasks, at different times and in different spaces, it consequently is a practice utterly dependent on a moral economy of complex bonds both within the lab and without. Today, the practice of genomics requires vast capital outlays, the recruitment and coordination of significant numbers of people doing varied and interdependent tasks, the establishment and enforcement of hierarchical organization, et cetera.

This world appears to be vastly different from that of Robert Boyle, with his domestic laboratory, and in many respects it is vastly different. However, as anthropologists observing the day-to-day practices at Celera Diagnostics, we can concur with Shapin's claim that "it is far from obvious that the world of familiarity, face-to-face interaction, and virtue is indeed lost. It seems quite likely that small, specialized communities of knowledge-makers share many of the resources for establishing and protecting truth that were current in the pre- and early modern society of gentlemen."[4] These resources include evaluation of the credibility of persons, machines, reagents, commercial arrangements, and the like. Although there are myriad formal social technologies of surveillance, evaluation, and legal remedy deployed in the world of contemporary big science, it is clear—if one gets close enough to look and to ask—that a complex, and thickly woven, web of mutual ethical evaluation and refined ranking of credibility functions everywhere in these microworlds.

Obviously, twenty-first-century genomics is no longer a world of gentlemen and domestics. Rabinow has characterized those who run organizations like Celera Diagnostics as "technocrats" in the positive sense the term is used in France, to refer to those who direct technicians, or as "technicians of general ideas," those who take grand concepts and make them into practical systems.[5] Perhaps it would be appropriate to update another originally premodern term: to characterize another body of people in organizations like Celera Diagnostics, "steward." The term originally designated a domestic role but came to have broader applications, from someone who directs the work of others, as in "shop steward," to someone associated directly with the safety and comfort of others, as in "airline steward" (and eventually "stewardess"), to one who in addition to their technical competence adds an ethical dimension of care, responsibility, and vigilance, as in "stewards of the environment." Bearing this complex semantic background in mind, we can accept a dictionary definition of steward as "one who actively directs affairs," as long as we remember the Greek and biblical roots of the term steward: *Epitropos* means "one to whose care or honor one has been entrusted, a curator, a guardian." *Oikonomos* is used to describe the functions of delegated responsibility, as in the parables of the laborers and the unjust steward.[6]

The roots of *oikonomos* are *oikos*, the household, and *nomos*, a term that is hard to translate but means approximately "law" or "norms." If we carry our etymological explorations just a little farther, we find that "economy" comes from *oikos* and *nomos* as well, although it evolved to mean the anonymous sphere of value bereft of the kind of care and personal tutelage embodied in the term "steward."

Man in the Middle: A Very High-Tech Steward

Joe Catanese is a steward in this corporate house of genomics. The obligatory passage point, or better, hub, through which data from all the discovery activities at Celera Diagnostics must pass is the high throughput lab. Although "high throughput" spaces are common now throughout the molecular biology industry, with its need for massive sequencing data of one sort or another, the Celera facility is distinctive in design, implementation, and power. The person who manages it, lives with it, nourishes it, cajoles it, drives it into higher and higher gear, adjusts it, readjusts it, despairs and revels at its performance is one Joseph Catanese. Joe speaks for himself in the interview that follows.

The high throughput lab at Celera Diagnostics is an impressive site for many reasons, reasons not obvious from a simple tour of the clean and light-filled modernist space, with its rows and rows of kinetic PCR machines—the Ravens—and its core robotic centerpiece. The distinctiveness is to be found in the practices that make the lab work. Its conceptual design exemplifies a form of inventive, improvisational ingenuity that brings together custom-designed, cutting-edge robotics, (customized) high-tech genotyping and expression machinery, and surprisingly down-to-earth equipment, such as a series of freestanding, large kitchen freezers. This is just one example of the contrasting elements that are juxtaposed in this space. In addition, due to a purposeful avoidance of complete automation—that is to say, automation for its own sake—technicians and scientists intimately cohabit with the machines and robots in the lab. There is an explicit value placed upon human work and judgment and a vigilant attentiveness to error, whether due to mechanical, informatic, or scientific problems.

Having had the option of automating practically everything, Celera Diagnostics' lab designers, headed by Joe but with input from many sources, decided not to. Celera wanted to avoid the possibility of all its research coming to a screeching halt because of a malfunction in one link in the automatic chain. Thus, a modular and flexible strategy seemed preferable, and this strategy fits Celera's multicentered overall strategy, the decision made at its inception to proceed on a number of disease association studies. But the avoidance of total automation also embodies or incorporates or instantiates the art of establishing ratios of reliability and judgment best suited to high throughput but uncharted procedures. All of Celera's plans for the high throughput facility are "in principle" feasible; moving from that state to de facto is Joe Catanese's mandate.

Balancing man and machine in high-volume studies comes with its own set of "challenges," to use the jargon of the business world. The work is highly monotonous but requires careful attention to detail. Problems may arise out of the seemingly inevitable neglect of detail that overconfidence in the regularity of machines often brings about or out of the less obvious, but apparently predictable, waxing and waning of human energy and attention. For experienced managers like Joe Catanese, charting, and thereby anticipating, the causes of human error comes to be almost second nature. Predicted high employee turnover rates mean that Catanese has to keep careful watch over employees, and make sure that they remain competent and interested at all times. Mistakes in this kind of environment are costly and may lead to the discarding of an entire day's work. Catanese, having gotten the lab up and running in record speed, now irons out kinks while keeping watch over research and results. A warm and honest character, he jokes that two years ago, when approached by John Sninsky and Tom White, he didn't have any gray hairs. Clearly, stress and pressure accompany Catanese's work, though he seems to enjoy what he does and to do it wholeheartedly. For the record, even now in 2003, Catanese has relatively few gray hairs.

Interview with Joe Catanese, February 24, 2003

PR: You are charged with an awesome amount of responsibility, designing and taking care of a truly innovative complex. We would like to know something about your background.

JC: Okay, I'm going to tell you exactly how I got here. It's a twisted path. I grew up in Niagara Falls, New York, graduated from a small Jesuit college, Canisisius, in Buffalo, with a bachelor's in biochemistry in 1978. During my senior year I was appointed head biochemistry laboratory instructor, and I was fully intending to apply to dental school. During that one year as head instructor, conducting classes and laboratories, I got the idea that perhaps I'd like a career in research. At that point, my professor offered me a job. He had close connections at a local cancer hospital, Roswell Park Cancer Institute, at that time the third largest comprehensive cancer center in the United States. One of the investigators was looking to hire a technician. I thought, Why don't I take a year to see if this is something that I really want to do, then continue on with my schooling? I began working right after school ended, and I absolutely fell in love with bench science. I was fairly good at it; I had the capability, at least back then, when I was twenty-one years old, of doing three, four, or five concurrent experiments and keeping them running, which was amazing for my boss at the time, who was probably experiencing then what I'm experiencing now, being in my mid-forties, and realizing that you can't quite remember everything. In hindsight, I understood what he did; he started throwing money at me, because he knew my plan was to stay for roughly a year and then move on. I was quickly promoted.

Most of the scientists were dealing with protein biochemistry at that point. I was doing some enzymology work, protein purification, assay design and development. We purified proteases from snake venom and used all the tools to interrogate the structure of plasma protease inhibitors. In any event, it wasn't long before my year was up. So to make a long story very short, I decided to stay and postpone my graduate studies. I spent the better part of about fifteen years there with one principal investigator [PI].

PR: And what did your parents think of this? And what did the Jesuits think of this?

JC: [*laughs*] Those are very good questions. My parents wanted me to do what I wanted to do. I come from a working-class family. My father was a U.S. mail carrier. My mother supplemented his income by cooking at various clubs. I have four other siblings, of which only one went on to college. So my parents were thrilled that I was out doing this work and I was making a fair income.

Now you asked about the Jesuits. Well, my head biochemistry professor, he was up in arms: he stated that he had trained me to do research, to take a year to think about it, and then go on to get a doctorate. "Why are you doing this?" I told him, "You had talked about the difficulties in having your doctorate, getting a job at a company where you are told what to work on, or in academia, which I now have become aware of, the only thing that you really get to do is bring in grant money." That is your goal in life, to bring in grant money, and the pressures are apparent. I questioned whether or not that was what I wanted for my life. Even at twenty-two years old, I had seen investigators who lost grants and the effect it had on their lives. It is a competitive environment, and it would have been rather naïve to think it wasn't. I sensed by doing this, I'd be able to do the experimental bench science that I wanted to do. I wouldn't have the pressure of having to go out and bring the money in to enable me to do it. I'll let somebody else worry about that, which carried me on for ten to fifteen years.

PR: Was your Jesuit advisor convinced?

JC: I can tell you this much, years later he said, "Why do you think I'm teaching at a small school in Buffalo? I went through the same thing with my advisor. He essentially said the same thing that I did." I look back upon that right now and I laugh, considering I'm putting in anywhere from eleven- to twelve-hour days, every day of the week. But at that point in my life, I worked an eight- to eight-and-a-half-hour day, I did my experiments, I was able to go home, I coached Little League baseball for ten years, before I had children. In any event, I stayed with Roswell Park for the better part of twenty-one years.

PR: Where were you at that point, in that hierarchy?

JC: When I left Roswell Park, I was a laboratory manager, and I managed a major effort. I had the total respect of my peers. I mean, people knew I didn't have my Ph.D., and people questioned why I didn't have it, but I don't think anybody ever questioned my abilities, which made me feel good. But, as grants are prone to do, they come and they go. My first boss lost his grants after thirteen, fourteen years. So, I tossed about for a year or two among a couple of other PIs, who were more than willing to pick me up. It was at that point, though, that I started questioning my earlier decision not to go to grad school. By then I'm married, with a kid, I'm getting into my mid-thirties or late thirties, and with the grants running out I don't know where I'm going to be in two months. Why did I do this? I got blindsided by money and responsibility, but I am wondering—maybe it wasn't the best for me in the long haul.

That's when I met Pieter DeJong, a superb technologist. He had been at Lawrence Livermore but came to Roswell to work on the making of bacterial artificial chromosomes (BACs). This was a new technology at this period of time, '94 to '95. It was also at the same time that the human genome effort was ramping up. There were problems of scale to make the large-scale sequencing work, and BACs seemed a promising way to improve things.

So, Peter joined Roswell. He was looking to build up his staff, and he was aware that I was going to be without a position. He approached me and he made me an offer, presented me with the first challenge that I had in a very long time, which really changed my life. The challenge was to think in terms of big science. The problem was that there was an incredibly complex technology, probably very similar to the first days of PCR, where a few very gifted and trained hands might be able to get it to work. Highly technical hands could get it to work, but the technology simply did not transfer readily into research laboratories. So it became obvious to me that we had to stop thinking about transferring the technology. Moreover, we had to think about becoming a center of excellence. We could make these BACs better than anybody in the world. Now we need to do it at a higher rate than anybody in the world. We needed to take a very difficult technology and do it more quickly and better, and then put these resources into the hands of many researchers. That was my goal. During this time, the idea emerged of forming a nonprofit company, called BAC PAC research. So—we're talking '96, I believe, '97 era—we start thinking about how we can make more libraries, make them faster and better, provide those libraries in terms of clones to researchers. There were many technical obstacles. It was challenging.

PR: Was this federally funded?

JC: Yes. The initial grant was through the U.S. Department of Energy. Pieter had worked at a national lab, Lawrence Livermore, and knew all the players. This funding was quickly followed by an NHGRI grant. We weren't looking to generate revenue. We were purely looking to sustain the operation.

PR: Did you ever seriously consider starting a company?

JC: Quite candidly, I had argued very early on that we should spin this off into a company. I didn't think that the company would have longevity, because technologies come and go. Very few technologies withstand twenty, thirty years. Always something better comes out. But I felt that we would be able to spin this off, and as the technology grew, we would be on the cutting edge driving that technology. Pieter felt it was very important that we provide this service in the least costly manner to researchers worldwide, and I understood his point of view. We also had to get into the legalities of spinning it off in terms of publicly funded, NIH, and Roswell Park interests in the state of New York. In any event, I was very happy to continue doing anything to grow, and we grew. And we became rather well known in the genomics community. For the public human-genome sequencing effort, over 80 percent of the sequence reads came from the library we created. So 80 percent or more of the data in the public databases came from the clones that we made and we distributed. We distributed those libraries to every major human-genome sequencing center, not only in the U.S., but also in France, in Germany, in Japan, in Britain.

PR: There are multiple advantages when almost all the labs are using the same material.

JC: Well, what ultimately happened, though, were the growing private efforts at Celera Genomics, where Craig Venter announced his plan to sequence the human genome in a matter of three years or less. There was tremendous pressure and public effort to move forward. We had to spread out the burden of producing the sequence. They just kept on going and going, using the first BAC library that we made. So we were doing rather well and that led to many more grants. About that time, in early '99, Pieter was acting director, but I guess things weren't going as well with the plans to build the department and its infrastructure. He was approached about moving to California to be director of molecular biology at a pharmaceutical company. Pieter approached me and essentially said that he wanted to accept but that he wouldn't accept without me. So here's this home boy who had lived his entire forty-two years in western New York, had one job for twenty-one years, and three children at the time, a wife who was a stay-at-home mom, whose parents lived in the area—all of our siblings, most of the family, lived in the area—and he's presented with this dilemma of packing up and moving out to Alameda, California.

PR: The American dream.

JC: Right. Okay. They flew us out a couple of weekends in early spring of '99 to help me make my decision. It was funny, because people in the area would say, "You're probably shocked at the prices of real estate." I said, "Are you kidding? We were electrocuted." But, you know, I thought back to when I was twenty-one years old and had become so comfortable with not wanting to challenge myself to maybe do as well as I could do by going on to get an advanced degree. I said, "No, this might be a once-in-a-lifetime opportunity, and I'll never know unless I do it." In Alameda, the houses we looked at were out of our price range. So we kept moving further and further away, until we could find something that is not a step down, with a good school system, and I could live with the commute. So we ended up in Concord. But it's probably the best move I made. I compensate by doing what I have always done: I come in real early in the morning. I leave my house about 5:30, so I don't hit much traffic. Going home is a little bit more problematic. It takes anywhere from forty-five minutes to an hour and fifteen minutes. But I've learned to live with that.

PR: What was the job you took in Alameda, and what was the research there?

JC: At that point I'm working for Parke-Davis/Warner-Lambert. Pieter had struck a deal, where 25 percent of our company time could be spent continuing the academic effort so that we could continue getting public grants. But that was when my misery actually began, because now I was trying to do two jobs. I'll spare you the details. The research there was really BAC library construction and distribution. We were really a technology-directed laboratory. So I was doing this dual role, back and forth on Route 880, which, quite frankly, presented problems for me. I'm coming into one site at about 6:00 in the morning; I stay there until about 9:00 in the morning. I then take highway 880 to Alameda; I stay here until about 5:30, wait out the rush hour. It was very difficult, but I continued to do that because I was committed.

Then Pfizer announced a hostile takeover of the company we were working for. So less than a year into this major career move, we are now Pfizer, which had a different company philosophy. The company went through a period of about a half a year of functional integration; they decided that the Alameda operation was redundant. So now, less than two years into this major move, I'm not going to have a job. I was okay; I handled it well, because I wasn't crazy about what I was doing—I was killing myself. I had been working for a company, Parke-Davis, that had never, ever lain off an employee; now Pfizer laid us off. I had the option of staying with Pfizer and moving to Ann Arbor, Michigan, or Connecticut. I couldn't do that to my children. I had just moved them two years earlier. I couldn't do that. Pieter arranged a job at Children's Hospital. I would run the effort there.

And then I remember a day in December of 2000: Pieter came in my office, with two guys following him; one was Tom White and one was John Sninsky. I had never met these guys. I had heard their names before. He introduced me and said they were interested in looking at the lab space that I had been building for Parke-Davis. I walked them over there. We were building the space for the molecular biology laboratories to do genotyping, at a much smaller scale than we're doing now and with a different technology. They asked a lot of very good questions, and I saw those guys knew what they were talking about. I didn't know who Tom White was at that time. I learned very quickly. I knew that he had worked for Roche Molecular Systems, and I knew names like Kary Mullis, Randy Saiki, through all of my PCR work. Anyway, that was that. John stopped by a few days later and started asking me what my interests were. What do I want to do? I sensed he was fishing a little, and I said, "You know, maybe I'll give you a copy of my CV." So I became intrigued, and I started wondering, "Well, just what do they plan on doing?"

PR: Tom hadn't told you anything yet?

JC: No. Now, at this point, it was announced that we were losing our jobs. The company is bringing in biotech companies who are recruiting, because remember this is 2000. Things are growing big-time. All these companies are coming in, and they have money, and they're promising. A lot of those companies aren't in business anymore. I had many offers from companies that I came very, very close to taking. Then Kathy, John, and Tom came and gave the overview. They were looking for people. I mean, they had to start this company, and all of a sudden they had some good people available. I knew John. He was almost drooling. I know, for instance, that John and Pieter were meeting frequently because Pieter was interested, too. He was deeply hurt when I made the decision that I felt that this was best for my family and me. I mean I was really charged up about the level of responsibility that they were going to give me. Everything had aligned perfectly. What could have been a disastrous situation for me professionally and economically, and even personally, for my family, turned into a situation where I'd be able to grow, working with people who I respect, for a boss who I regard as a friend and who I respect more than anybody. So, I said, "Let's do it." I told Pieter.

PR: He didn't get an offer?

JC: Right. And I had decided that on February 28 I would cut my ties and start here on March 1, 2001. And so from that point forward, I had a single goal: to build this facility, to get it to run at capacity. And that's what we have been doing.

PR: What kind of details did Tom, John, and Kathy have two years ago? Did they know what they wanted really from you?

JC: Yeah, sure. I was given rather broad guidelines in terms of the magnitude of studies that we would want to be able to perform on a yearly basis. I viewed this challenge much in the same way I viewed things back in the mid-nineties. You try to take a technology that has been, up to this point, rather artsy and try to get it to run in an industry-research-scale factory. I felt I had some ideas about how to go about doing that, because at that point in time we didn't know exactly how we were going to do the experiments, so it was clear that I needed to build a flexible platform. So I immediately came to the conclusion that I needed to build a modular system, as opposed to an integrated platform. I had had problems with integrated platforms in the past, with bottlenecks causing everything to slow down. I simply didn't know the right technology back in early 2001.

PR: Did anybody?

JC: No, no.

PR: Celera Genomics didn't, and ABI didn't.

JC: It was a totally different technology. They had a process to do high-throughput plasmid sequencing. Our technology was totally different.

PR: Was anybody doing anything like what you're doing here two years ago?

JC: On a real-time PCR platform? Absolutely not, at the scale we proposed. Other facilities were getting started with high throughput genotyping, but the technology varied from facility to facility.

PR: So they trusted that you had the mastery of existing technology, but they also made a judgment that you also were the kind of guy who was flexible enough to take big risks and to try new ways of doing things. You were not just a tech man; you were a big-scale manager too. You were the type of guy who would be comfortable with the scientific input that was going to be cascaded in, is that right?

JC: That's partially right. There are many people in this organization that know much more about the details of real-time PCR—in primer design, in polymerase characteristics, in terms of assay conditions—than I do. No doubt about that. But I've always prided myself on the fact that you need to know the underlying scientific principles before you can even think about automating a process, because it's really the robustness of a scientific principle that's going to give you the capability to design and maintain a system.

PR: Of course, you're a modest guy. Kathy said the other day that the high throughput lab is absolutely the linchpin in the whole operation. Other people are also doing important things, but without this, it doesn't work.

JC: True. I tell my people quite honestly that we are the engines that are going to drive this company. I told them from day one, when we were recruiting, through every weekly meeting, that when you sputter, this company is going to sputter. It is a major responsibility, but I have a lot of help from my colleagues. John was essentially telling me, "Tell me how you think that we are going to do this." At first I saw the challenge in a rather naïve fashion. I wasn't afraid; I don't know why. I already had an idea in my mind of what types of instruments I would probably use to do it, what types of people, the number of people I would need.

PR: How long did that take?

JC: About three or four months. I think by early summer of 2001 I was interviewing most of the major players who were handling robotics. I had a pretty good idea what direction we were going to go in, but I felt we had to do our work diligently and bring in all the major players, present them our problem, and then allow them to present their answers to our problem. Three or four companies presented options. The one that I thought that we were going to go with, they said, "You know what? We can't help you here. We simply can't do what you want to do, with the volume and speed you want to do it." I ultimately decided that I had to go with a company that was flexible enough to allow me to change my mind midcourse, because I really didn't know how we would get the DNA and all the primers put together. A lot of it was done on the fly, going down to the factory and saying, "Now this isn't going to work." The way in which I envisioned the actual workflow changed many times. So it was cheery coming in every day, identifying the problem, and then tackling the problem. Sometimes we would tackle it in a day, sometimes we'd tackle it in a month, but we never lost sight of the fact that, "Okay, we have a deadline: we want to move into this facility in September of 2001. We have to have our plan ready. By December we want to be prototyping on the robot. We want to be taking delivery by January (and then it ended up being March). We want to be operational so that we can complete the study by the end of the fiscal year, by the end of 2001." As a company, we had never missed deadlines. The company plan was predicated on the fact that you were going to be able to make this happen, and then you had to worry about other things. My responsibility was, whatever Tom, John, and Kathy came up with, to build a platform that could handle it.

I hired a lot of entry-level people. We now have fourteen people in the group. I hired about six people and followed up with another two or three people in mid-2001. We still didn't have the facility. We placed them in Disease Area Research groups for a number of months to allow them to become familiar with the technology and to get their feet wet. This laboratory is run basically on the back of the people who move the plates back and forth. I work under the premise that I will have the highest turnover rate of any department in this company. People tend to get fatigued. People tend to get frustrated, and unfortunately within our group we don't have clear areas for career development. This is not to say that they can't go into other areas of the company as the company grows. I'm happy to say that we've had zero turnovers, whereas other groups have had turnover. The employees in the lab do a tremendous job. They understand our role, our importance in this company. So I was very lucky to be able to attract young people who want to work, understand their role in the company, and come in to work hard every day.

PR: How did you recruit them?

JC: It was done over two interviews where we interviewed quite a few people—maybe twenty people, maybe more—pared that down, talked to them again. We made offers.

PR: Were you looking for character or specific skills?

JC: I'm not looking for the guy or gal who had a 3.5, 4.0, who had a publication already. I'm looking for the right person, who I think has the personality and mindset that could be able to handle doing this level of scientific research. It is so totally different than what they probably envisioned they would ever be doing in their careers. You have this romanticized view that you'll be working in a laboratory, have a hypothesis, do a little experiment or two. You look at the result, and then you base your next course of action on the result. We don't do that. We don't do that at all. We do the same thing over and over and over again. We do it as well as we possibly can do it every time. The most important thing for me was to find people who, through the interview process, I felt were going to be able to handle the stress of doing this work.

I meet with John on a daily basis and I tell him what my concerns are. We've made a number of statements, both externally but mainly internally, in terms of what we want to do, in terms of the number of studies, for so many disease indications, so many markers per study and in what time frame those are to be done.

Now that we're up and we're running, I know what our well capacity per day is. I know that we can do approximately 145,000 titre plate wells a day. I know that that multiplies up to over 600,000 a week. And I know how many wells a study will take for 20,000 SNPs, 25,000 SNPs. So I know how many studies we can do a year if we dedicated 100 percent of the capacity only to that cause. But I also know that we need to keep on building assays, validating assays. I also know that in addition to our pool of discovery work, we need to take those hits and do individual genotyping to confirm that. I also know that when we do find a hit in a sample population, we need to bring in another replication sample population and replicate that before we go public with anything. So we can't be doing 100 percent discovery. I know from prior experience that you cannot build a platform and expect it to run at 100 percent.

PR: Does that mean that you and the company are closing in on what would be the first complete cycle, let's say Alzheimer's? The vision is now coming, with seemingly exciting results—to the end of proving it can be done?

JC: Yeah, I believe we will be there once we do a whole genome scan on pooled case-control DNA, where we have no real up-front knowledge of genotypes. We pool samples, we find markers that hit in these case pools at a higher or lower frequency than the controls, and then we break out those hits for individual genotyping. Once we do that cycle, then it's come all the way around; the proof of principle has been established. Then we just do more of it. We strongly believe that we're not going to find that magic SNP, the one SNP. There's obviously something else out there, possibly operating in combination with environmental factors. We truly believe it's going to be, as Tom says, a constellation of markers. We have hundreds of markers that show significance that can ultimately be a part of the panel. But we believe that we'll be able to distill that down to a constellation, take twenty, forty, fifty or a hundred markers. That constellation will become a true diagnostic panel that's going to provide relevant medical information, not only to a provider but also to the individual patient, so that they can make medical decisions.

PR: Last question today, what could go wrong?

JC: We might not find anything. We're finding markers in individual genotyping. That's not unexpected. Well, it's good news. Science makes sense. You have a single point mutation that changes an amino acid of a particular protein. Some of these could be disruptive to activity. And this is what people have been finding for years and years; anybody can do that. The key is doing it at scale such as you can afford to have misses, because we won't have a hit every time we conduct a study. What we need to do is have a few singles every now and then, a double here or there, triple or home run once a year, I don't know. I know that we've done simulation studies on what we discovered on Alzheimer's and asked the question, "If we did this by pools, would we have found this?" The answer is "yes." I'm very optimistic but I'm also very practical. I want to do it in a lab, and I want to be able to say, "Here are our data, we found it in the pool."

The Machinery: Interview with Joe Catanese, March 7, 2003

PR: Let's review what you presented in previous interviews: What existed, and how did you put it together?

JC: I joined the company March 1 of 2001, so I've been here two years now. Two years and a week, and I didn't have any gray hair when I began. And that's true! [*laughs*] Almost, almost true. John presented me with the challenge of building a laboratory and a laboratory process that would enable us to do between five and ten whole-genome scans per calendar year for disease association, using SNPs as markers. And to do, perhaps, three to five messenger-RNA expression studies—again, using the same automation platform and technology within that same calendar year. And do some level of what we call ribotyping: looking at differential messenger RNA expression from each of the two copies of every gene—again, within the same calendar year on the same platform.

So that was the basis for sitting down and trying to figure out how we're going to do this. And as I explained last time we spoke, we spent the better part of the first three to five months trying to conceptualize the plan. At that time, we were building the space and talking to most of the leading suppliers of liquid-handling robotics to see what we could purchase off the shelf and do minimal modification to, so that we could be up and running ASAP. We also wanted to place the burden of maintenance and development on a secondary supplier, as opposed to trying to build a custom system in-house, which would require engineering expertise and perhaps some assembly or manufacturing expertise. We quickly came to the decision that what we wanted to do was build a modular, automated platform. By modular what I mean is that the various tasks that lead up to the product are separated in a logical fashion so they can operate independently of the entire process. In the event that there is a problem at any point in the process, that problem doesn't affect work at the other areas of the process. We conceptualized how we were going to do it, in terms of moving liquid around in tubes and plates and what magnitude that would be and quickly came to the determination that we were going to do this at a level that no one has attempted to do before in this particular type of format. And that's not to say that there aren't other organizations that are doing the same level of genotyping or expression work, because there are many organizations doing much higher levels of expression work, but they're doing that on a microchip platform. We're working in a wet environment. We're putting liquid into a container. In this case, it's a 384-well plate and that presents unique challenges.

PR: This is a very big fork in the road. Roche and others are really going somewhere else technologically?

JC: The core technology is totally different. And therein lies a lot of what we feel is our advantage. We believe we have much more sensitivity relative to these other technologies, but it's a challenge to do it at the level that we want to do it. This technology was really done in what I would call a one-off type environment where particular assays were hand-curated in terms of design and then modified many times to get them to work optimally in an assay condition; then multiassay conditions were employed for the handful of assays that we had in order to get all of them to work as well as possible. We started with 15,000 assays. That number grew to 20,000. It soon grew to about 35,000 assays. Right now, we're aiming at about 20,000 for our scan that is going to begin in late April, but that number, I fully expect, will grow to 30 or 40,000. And the challenge is to get every one of those assays to operate under a single operating condition. John formed two teams among disease area personnel and asked them to look at various enzymes and various conditions to run the assays in, so that the organization could come to a determination about what enzyme to use and what assay conditions would be optimal for that enzyme. We did "window studies" looking at various concentrations of every component, and we came to a final operating condition that we felt comfortable with.

PR: Let me just gloss this. Molecular biology, in a very short period of time, has gone from an essentially craft activity to a type of industrialization but an industrialization that can accommodate craft: these are individual assays specifically crafted for a particular experiment. The next step is very large-scale industrial activity that requires a radically different scale of standardization. The microarray chip, in its own way, was large-scale but was an instrument that can just do more and more of the same thing.

JC: Exactly. That was the challenge: to find a condition and then to run that condition time in, time out, in an accurate and precise manner. In other words, a reproducible manner. So we soon decided, after interviewing the major players in robotics, what way we wanted to go, and we chose two different manufacturers to provide us with our platform. We chose the first one, Beckman/Coulter, simply because at that point in time, and I still believe this to be true today, they had the robotic platform that presented the highest degree of flexibility. At this point, I still didn't know exactly how, to the minute detail, we were going to do this. Just how are we going to bring this DNA into a reaction? Just how are we going to take these primers and get them into a format for a reaction? Not knowing that, you can't build a robotic platform for a particular exercise because you don't know your exercise.

PR: You needed to be prepared to deal with potential variability from the different disease groups?

JC: The variability would come in purely from what we found in the disease area. How many microliters of a reaction are we going to have to run? That would really direct me in terms of how I'd have to do the experiment. So I needed to have a platform that gave me the flexibility of perhaps moving one component around at a time, or moving 384 components around at a time. Perhaps being able to pipette a half a microliter or perhaps 150 microliters.

PR: So there is a truly massive difference of scale?

JC: And we're still operating at that level today. We use an instrument, and many instruments can do this, to do what we call single-channel pipetting, one at a time, or as many as eight at a time using one particular device. And another device can do 96 wells at a time or, if we choose, 384 at a time. So we decided on that platform, and currently we have eleven of these robots and we're soon to purchase our twelfth.

PR: How much do they cost? Ballpark?

JC: About $200K apiece. Now these robots wouldn't solve our ultimate question: How are we going to assemble these 384-well PCR trays per day? Because at that time, we thought it was going to be somewhere around 250 to 300 trays, and today we're beating that goal. These robots were not going to get that done for us, not at all. I was familiar with another robotic platform from my experiences with making BAC libraries that was based on magnetic beads. We worked with Packard Bio-Science to build what is called a DNA plate track robot—massive, long robots that Packard supplies modules for, and then you put these modules together in a custom application for whatever your task is—that would use this platform. These little plates are moved across an assembly line and are being addressed at various stations along the way. So I saw this as a way to do this reaction, because I could envision where we can take out the DNA that is made by the DNA lab, the primers that we made on the Beckman FX robot in the primer lab, and bring them out and combine them as we move along an assembly line, and cross various liquid-handling modules that grab the contents of one and grab the contents of the other and combine it into another one and stack these up into a new stack and allow us to move a vast amount of plates through in a minimal amount of time. That was really the challenge. Nobody had tried to put this many reactions together in that period of time—in an eight-hour day, about three hundred to four hundred plates. So we brought Packard in, we presented the challenge to them, and working together, we came up with a design. We placed the order in September of 2001. The robot was built over the next five-month period, approximately, down in Southern California. When we visited the factory, late January of 2002, to do what was called "factory acceptance," a couple of things became obvious to me. It was going to work, no doubt in my mind, but we needed to do some major modification to the way in which it was currently working. So the robot went through a little bit more fine-tuning in the programming, and then it was shipped up here in the first week of March 2002.

PR: Is any of this now a secret?

JC: We don't want to advertise two things: that we're pooling DNA, although that's becoming common knowledge, and how we're doing PCR. We're going after the functional markers, functional SNPs resulting in an amino acid change. So we basically don't have many existing reagents available to us. Right now we are kind of very low-profile on all of this, as you're probably well aware.

PR: Academic critics have a view that the main issue is patenting the genome—greedy capitalists taking people's deepest secrets, putting them in deep freezers, and all of that. However, it seems to me that the real game is timing. How long do you need to keep this under wraps? We're talking about quite specific periods of time. But whenever this really goes public, then you're proud of what you've done?

JC: Of course! And our plan is not only the normal press release mechanisms that most companies use but we fully intend to publish the discoveries that we're making in peer-review, scientific journals. Of course, we will establish intellectual property [IP] on those particular markers at that point. Our role here as scientists is to develop products for the company. And in order to do that, you have to have some degree of confidentiality early on so that you can establish a claim. And then, as scientists, we need to communicate that to the scientific community, in order to better mankind. We fully hope that others will follow up on the discoveries we make.

PR: So there are periods of confidentiality and periods of publicity?

JC: I would argue that this is no different than what exists in the academic world. I spent twenty-one years in an academic lab, and when we were on to something very hot, we did not publicize it until we submitted a publication! Once you had the submittal date, then you talked about it publicly, because that gave the link to your work that justified writing grants.

PR: Okay. You've gone from big concept to real machine work to modification of what's available to your specifications. The truck arrives with the machines . . .

JC: So we bring it in. That particular big robot that you've seen is actually two instruments. There's the robot itself, and then there's an environmental chamber outside of it that was contracted as a custom job from another company. We had to remove the glass windows in the large conference room because that was all built on one frame. It does not come apart, and we had to have two forklifts lift it up carefully and one push it into the window, and then we moved it into the lab. We have pictures! I didn't even want to watch! I said, "Tell me when it's done!" Anyways, we get it into the lab, we set it all up, and we spent the better part of two months changing the methods that were programmed at the factory, to give us the accuracy and precision pipetting that we felt we needed. These were mainly software modifications.

    "We" is not I, and I want to make that very clear. Not me! If you asked me right now to go to that instrument and get it to run, I couldn't do it. That's one of my frustrations: I've always been at the bench, and since I joined the organization, I joined with the understanding that I would be a working manager, and that's never going to happen. I won't get into the lab. I've been in the lab maybe three days in two years. "We" is a team of people. For example, David Hung was a young person who joined us from Affymetrix. We didn't know exactly what he was going to do here, but he was talented and we figured we'd find something for him to do—and we did! We gave him the responsibility of being the chief programmer of the plate track robot. I wanted variance of less than 2 percent in pipetting; Packard would guarantee less than 10 percent. When it got to us, it was in the 5–6 percent range. That was not going to work for the particular technology that we needed. David spent a good deal of time—very, very, very productive time—to modify the way in which the robot aspirated liquid and dispensed liquid to improve the accuracy and precision. With the new setting for accuracy and precision, David was able to generate about fifty plates in about one hour. Now we have another dilemma: we now can do it the way we need to do it technically, but we're not going to be able to achieve the capacity we planned for. In a normal eight-, nine-, ten-hour day, we are not going to be able to generate these three hundred, four hundred plates. So the next challenge was to take all of those modifications that he made, keep the precision and accuracy, but shorten the time frame, and that's been an ongoing project. It probably came to completion, I would say, about two months ago. We're now at the point where we have less than 2 percent error and we can now do forty-eight plates in about fifty minutes. And this is just from analyzing and trying and getting better accustomed to how this instrument works. That is purely David; I'm indebted to him, because now we're at a point where we can do it at the accuracy we need, and we can do it at the speed we need to meet our goals.

PR: So you're working under intense time pressure but with a certain amount of lead time because the tissues and samples are being collected during this period of time. So everyone else is ramping up too. What month are we in now in the recounting?

JC: We're getting toward the end of June of 2002. One of the goals that we had set for ourselves for fiscal year 2002, which ended June 31, 2002, was to have completed a study. We wanted very much to meet that goal, so we embarked toward the middle part of June on the Alzheimer's study. This required the arrival of some DNA provided by a collaborator to be entered into our database system, then to be extracted, to be quantitated, to be standardized to an operating concentration, and then to be put onto the plate track, et cetera.

Now the assays were not as much of a problem because we had been building assays from early February 2002, so we had a half-a-year head start on that. The disease group's responsibility was to pick the assays they wanted to run first and feed that information to me. The goal was to put all the samples, roughly about 1,000 samples, through a panel of

96 markers. Now keep in mind that our goal now for a study is 20,000 or 25,000 markers. But still that was a big undertaking, because most academic labs struggle to put one marker through all their samples in a week. But it was the lead-up to getting all the modules in place so that we could make everything flow into the plate track and produce the data. And once we produced the data, the other side of the operation was to grab that data, process that data, and make it visual so that I could make sense of it from a quality point of view, the disease groups could make sense of it from a finer scientific point of view, and our statistical genetics team could take that data and analyze it so that by June 31 we had a report. We made it!

PR: So this is a two-hundred-fold scaling up of what an academic lab would be doing. Is that right?

JC: Yeah. Let's say one-hundred-fold scaling up. Look, I know, in other labs, when I would be doing one marker across a thousand samples, it would take me about four to five days to do.

PR: And everything worked?

JC: Yes! We did a study, but we did not find a significant marker among those 96 markers, which makes total sense. I mean, we would have to be very lucky. But then we continued along the path of doing the Alzheimer's study. I know Paul's aware that we have results. We continue with 96 markers at a time, but the frequency in which we can do that becomes shorter and shorter and shorter, to the point that we are now doing well over 2,000 markers in that particular study. And that, I should point out, has all been done using individual DNA. Our business plan, our scientific plan, is to pool the DNA and that's where the power of our technology, the power to do the four to six whole genome scans per year and to follow them up and replicate them comes from. Then we plan to follow it up with individual genotyping only on those markers found to be significant, which drastically reduces the cost of individual genotyping. We'll do that on pools of DNA, replicate it on another case-control population, or cohort, as we call it, and then bring those markers down from 20,000 to perhaps less than 100 markers. Those we can then run and confirm for a third time on individual DNA. We can do that in a day. So our ability to look at the whole genome in a study increases our chances of finding as many markers as possible associated with a particular indication. That is why we need to build a system that enables us to do this in a timely and cost-effective manner.

Sometimes the machines get very tired, and that's why we're constantly calibrating them, checking them out. Keep in mind that these machines, they're new too. This is not a standard technology that has existed in an academic lab for the last twenty years. These are the latest generation of robotics, and every time they build a new generation, they're obviously trying to push the envelope a little further in terms of the speed—hopefully, in terms of the accuracy also. But it normally involves making parts smaller and smaller, and these get pushed out to the market very quickly. As a matter of fact, we are one of the first laboratories to receive some of these robots. We're beta test sites on some of them, so we find problems with them that perhaps many other laboratories don't find, purely because here the robots are running constantly. But once we established where we needed to be, it was our responsibility to get ourselves there. The manufacturer only takes you so far.

PR: Okay, so we're in July 2002.

JC: We're in July, and now we are operational.

PR: I hope you celebrated.

JC: Yeah, we did have a celebration. We had a birthday party. I mean, our "born" date! Kathy decided our birthday would be on July something or other; I can't remember what the day was. But we celebrated in the sense that we were able to accomplish something—a small thing compared to what we wanted to accomplish overall. So between July of 2002 and where we are now, we are simply getting better at what we're doing. That comes from repetition. That comes from many, many modifications of the LIMS.

PR: At this point, obviously, stress is high. Is there conflict as well? Or is that a distinction that means anything?

JC: Minimal . . . minimal conflict. Maybe I experienced stress back then, but it pales in comparison to the stress now. I guess that I never had any doubt that I could build this. I knew I could build it. But you cross a threshold from building to operating. Now is kind of the proof of the pudding. You built this, and now will it work? And that's where the stress actually begins. Conceptually it was all there, but often concepts don't translate to practices as well as you had hoped. All of these little minute problems that on their own you would think, "Well, that's solvable within an hour or a two-hour time frame," start to compound and hit one after another. So you feel like you're constantly bailing water out of the ship, and you can keep up now, but how long before you collapse? Then you're in trouble. I feel like that now, because the workload continues to increase but the capacity has now been met. We're into a new kind of a juggling situation. But in terms of stress? I could tell you this much: I've been in this game for too long now, and I've had other positions where I dreaded getting up and going to work. Never here! Never here! I go to bed thinking about this. I know I dream about it! And I wake up thinking about it, and it doesn't stop when I get in my car at 5:30. Doing this turns me on. There are times when I want to throw things. Every morning I move the data. It's about 6:20 in the morning, and there's only one other individual, Wally Laird, who comes in early. He comes in to see the work of the previous day. So I'm in the back corner of the lab and this machine has totally goofed up and . . . I take that back: we goofed up! Machines only do what you tell them to do. And so I have to admit it; I am swearing a blue streak because I was mad, and I just wanted to pick up something and throw it at the machine. I look out of the corner of my eye, and there's Wally, watching. I would be lying to you if I said that there are not days I go home just totally exhausted. The first thing that I want to do is, you know, just, go out for a walk, take a shower, get something to eat, and go to bed.

PR: Is there any interpersonal strife within the company?

JC: We have an infrastructure committee involving about four or five key people, disease area teams. We argue! [*laughs*] We argue back and forth, but the strength of this company is the way in which people pull together as a team to try to achieve the common goal. People understand what's on the line here. I mean, this is an opportunity that is once in a lifetime. We've made statements of what we're going to do, and now our reputations are at stake. I feel a great deal of responsibility to John to make it happen. I owe it to him! He trusted me, and now I have to do it for him.

It's not unfair to say that there are people in the organization that disagree with some of the choices that we've made, and that's healthy. That's very healthy! So we have very lively discussions, and we come to a consensus agreement wherever possible, and if not, John or Tom will make the call. We can't be paralyzed here! We know that's our enemy:

paralysis. We have to make a call. We have to move forward. Now, time may tell whether we made the right or wrong call, but I maintain that, as a company, I know we cannot sit around for three months and discuss how we're going to do something. I maintain we should start on the first month, collect the data, and let the data drive where we go, and this is the way we're going to do it. I told John we need to strike a balance between doing the perfect first experiment and doing a first experiment in a time frame that we set for ourselves. There's no doubt in my mind that this first experiment that we do will not be perfect, but I know three things: we're going to get a lot of data from it, and we're going to find significant markers, and that first experiment is going to guide us along the path of maybe one day trying to achieve a near-perfect experiment.

PR: Isn't that what an experiment is?

JC: I would like to think so. We're moving forward on Alzheimer's. I can't recall the date when the analysis pointed us toward the first significant marker. It was somewhere in the July–August time frame. These markers all run into one another after a while for me. At that point, we started to bring in the second study, a cardiovascular disease indication, collaboration with the University of California, San Francisco, and we embarked upon that, I think, in the October–November time frame.

PR: So for three months during the summer and the early fall, you were working on Alzheimer's?

JC: Yes, we were also learning from the data we were producing, in terms of what we had to do. We were refining the techniques; we were continuing to build the LIMS for the primer lab that's going to better support the applications that we're now doing; and we were still operating in the DNA lab in what I call a "manual fashion"—no LIMS support and hence a lot of manual record keeping. When I say manual, it was going into Excel spreadsheets and being manipulated there, printed out, and double-checked. Automation is now being used in the DNA lab, at the plate track, in the primer lab. All of this is ongoing as we continue to build the facility.

In doing this work we have also gotten a better knowledge base on our instrument platform. That's the Applied Biosystems system 7900 HT. We have forty-five of them on the floor out there. These instruments were developed for a research laboratory. No one has forty-five. Perhaps some laboratories have three or four. We are in the unique position of working across many instruments, looking at the reproducibility across many instruments, and running these instruments 94 percent of a twenty-four-hour calendar day. We're finding things with these instruments in terms of wear and tear. Our colleagues across the Bay would like to think that the machines fully meet the product specifications. We are probably stretching that a few orders of magnitude. So it became obvious to us that we were going to have to look at the platform to try and identify particular performance issues and whether or not we could make improvements to them. We are currently doing that.

It's a major thrust right now as we embark across the threshold to pooling. We've already done one or two pooling studies, very small-scale—192 markers, about 1,000 samples. We proved the principle, and it worked great, but in terms of machine performance, that's a frustration that still exists. But we've gotten great cooperation from Applied Biosystems. We have two people from AB here roughly every day of the week. We've now narrowed in on what we think the problem could be, and we're now trying to find out what's causing the problem. It's a very, very fine point—what we call plate effect: getting a different data point depending on where you are in the plate. We want to be able to either eliminate it or to figure out how we can control it internally. And there's a great synergy here, because we now know what to anticipate in wear and tear, and we've worked at preventative maintenance. Malfunctions are terrible, because we lose time and we lose data. So we've built a preventative maintenance program where particular parts that we have diagnosed as wear parts AB now incorporates into their maintenance schedule and, more importantly, designs better parts that don't wear. They probably didn't envision that somebody out there would be running the machine around the clock. So that's been working out very well.

PR: Can you imagine, Joe, that say in x months, you'll be sitting here twiddling your thumbs because you have solved all the machine problems?

JC: That would be a dream. But it's not in my dreams, because I've done this before. We are right now operating at near 100 percent capacity. Now, it's probably not as smooth as I'd want it to be, so we're going to make modifications to the LIMS, to enable the process to run a little bit smoother. But at that point, when you achieve a smooth operation, then the challenge is to increase your capacity. The best thing for us is that we are getting hit after hit after hit after hit, and things are going great, and we can say—or at least, this is my opinion—"This is working."

Quite honestly, my biggest challenge for the future is keeping everybody happy. This is an incredibly high throughput operation, and by definition that means very repetitive by nature. We hire young people right out of school. They are very energetic. We think we do a good job in explaining to them their role in the company in terms of being the fuel that runs the engine, that essentially runs the company, but I know we're going to have some problems with turnover as employees simply become bored of moving plates from point A to point B, of taking primer racks and moving them through the process. They want to move on in their career. They've done their period in the trenches, so to speak, and it's time to move on. And so trying to keep a core staff happy and then integrating new people and turning people over and not dropping off in capacity and quality during that transition process is always going to be a difficult task, so I don't think I'll ever be able to sit back and, you know, put my feet up and just have to show up and press the button.

About a month or two ago, we went through what I had told John we would go through, a six-month slump. After about a half a year, experience had told me, all of a sudden you would get a spike in error. And sure enough, it hit! We had a week when, if it could go wrong, it went wrong! The trick is to try to anticipate it and prevent it, but it's almost impossible. It is part of the challenge of trying to anticipate, of being proactive, of keeping morale up.

PR: Talia and I were talking about diversity issues. Has that come up at all?

JC: How are you defining diversity?

PR: Well, the usual American issues. You don't have a company hiring policy on that?

JC: Boy, I tell you what—I know I personally don't have one! What I look for is the best person for the job, and I couldn't care less about gender or ethnic background. I mean, you might see that we're probably evenly split male/female. And ethnic background, I really don't even know. I don't think about that kind of stuff. It's really funny that I don't think about that, because pictures of different ethnic backgrounds surround me, you know, and it's what we do. But no, that is not the basis of how we hire people.

One mistake can wreak havoc. When I was walking down from upstairs just now, I had just spent an hour trying to figure out how a mistake had been made, and it finally boiled down to this: Somebody made a guess that, "Well, this isn't the right number, so this must be the right number," and ran it through the system—and it was wrong! That one little mistake resulted in 1,250 DNA samples being run on the wrong assay. That's 1,250 DNA samples that we can never get back. It's gone! And what we have to show for that is duplicate data, good reproducibility. These are the challenges that are always going to be here.

PR: Do you take the person aside privately and talk to them?

JC: When I think an egregious error occurred just out of lack of attention, I'll pull that person aside and point out what was done, ask why the particular decision was made—because maybe I'm not understanding the thought process as to why that was done—and then point out how, perhaps, I would have done it. I'll also point out the fact that, when in doubt, come and see me. At our weekly lab meeting, I'm constantly reinforcing the fact that our role in this company is attention to detail.

The philosophy here is, Don't automate what a human being can do better and faster. So you won't see any automation in our platform for moving a plate from one robot to another robot. We could have built that, but it would break down all of the time. It's not asking somebody a lot to grab one stack of plates and move it over every half an hour. They can do it faster and they can do it better than a machine. A fully automated platform still wouldn't be running probably, because it would have taken us probably two years to build it. But I have to ask the question, Would we be further along in the process if we did it this way or the other way? And there's no doubt we've done a tremendous amount of work. Are we where we need to be? No. But I think I was hired to get it up and operating and then perfect it on the fly, and that's what we're doing.

PR: Tom and John are very proud of that. That's their vision of things all the way through.

JC: Well, they're the driving force. I mean, it's not uncommon for me to see John many, many times a day, but every now and then Tom will stop by and say, "So when are we going to start that particular study?" And that's Tom's way of saying, "Joe, we've got to get this moving," and that's never lost upon me. And that's a very effective management tool. I use it on my people.

Interview with Joe Catanese, April 11, 2003

PR: Joe, how are things going?

JC: Over the past month or so since we last talked, we've been, as an organization (high throughput group, the disease area, individual comp bio, statistical genetics), trying to come to an understanding of how we're going to go about doing the first whole-genome scan, and we've made decisions so far in terms of which assays in our collection of validated assays will be the first to be run. What we've chosen to do is to take both a practical and perhaps a conservative approach, in that we're taking those assays among our collection in which we have the most confidence. We have come to a total of nearly 10,000 assays in the first cut, and we've now loaded this list into our machine, and our primer lab is actively going about the work of finding those assays among our 350,000 primer tubes in our collection and bringing them together into what we call whole-genome scan racks.

So we're going about the work of bringing these assays down from about 3,500 racks of 96 to a set up of 90 to 100. We're really condensing them down, and at the same time we're beginning to generate the actual working primer racks that will feed the plate track to do the whole-genome scan. Quite candidly, at this point, we're behind schedule. We've run into a number of difficulties, some of which stem from equipment availability. Unfortunately, this all hit at the end of a quarter, and companies are very eager to sell their equipment at the end of the quarter, to increase their bottom line. The salesmen increase their commissions, so they parse out equipment based on first orders in and, unfortunately, we were not the first orders in for a piece of equipment. I'm encouraged that, over the next week or so, we're going to get these key pieces of equipment for the primer lab that will speed up the process. Also, we're not simply shutting down operation and saying, "Well, we're going to do a whole-genome scan at the end of the month, and so let's put all our effort into that." We're continuing to do our assay validation, our individual genotyping studies for Alzheimer's, our other studies; we've done exploratory work for the rheumatoid arthritis samples, and we've done exploratory work with additional samples from UCSF. Our plate is very full.

PR: So you're doing something that's never been done before, in addition to all the things that have never been done before? There's no stress involved?

JC: There's absolutely no stress involved in this job! And last week we had a virus infection, and now the database is down. I told you guys before, I love coming into work because there's always something to do. The day goes by very quickly, but the past couple of days have been quite a challenge. All that being said, there's no doubt in my mind that we will begin the whole-genome scan on April 28. We have essentially two weeks. We need to bring together a few more pieces. Everyone is very, very busy right now trying to put together their stratification of their samples into pool sets.

We intend to do two studies in the first whole-genome scan. The primary one will be rheumatoid arthritis. The secondary one will be a CVD [cardiovascular disease] study. That one will occupy less real estate on the assay plate in terms of the number of wells, the number of pools, and that group is trying to come to the final pooling stratification for the scan. All of this needs to come together by the end of next week so that our DNA laboratory and our high throughput lab can do the work.

PR: Can you describe that process for us?

JC: Sure. Embarking will not be quite as glamorous as perhaps people would like to believe. It will be much the same as what we're doing right now. We've built the first whole-genome scan essentially around the model of how we're doing the individual genotyping right now. That one plate of DNA will contain all the case and control pools for the two experimental studies, plus the appropriate experimental controls. What will really differ in this method is that, for the current studies, we essentially have ninety-six assays at a time and we're passing them across a thousand samples of DNA. So we're reusing the same assays over and over again. We won't be doing that for the whole-genome scan. We'll use them once; it's going to be a challenge for the plate track team because now, instead of dealing with forty-eight primer plates all day long, they need to deal with seven sets of forty-eight different plates. What's difficult about that? Just the logistics. They need to locate them, they need to remove the heat seal from them, they need to get them queued up, they need to get the barcode logged in, and they need to be doing this while the plate track is running the previous set. We are looking at increasing the amount of people to help out. Keep in mind that we do all of this assembly with two people. I think that's a testimony to the platform that we built, but it's more of a testimony to the two people who are doing the work, and that should never be lost. These people are occupying the entry-level positions in our organization. They are hired out of school, and they attack this job with an attitude that really should be commended because they really are the true heroes of high throughput, and I'll say that over and over again.

PR: We believe you.

JC: Yeah, they are the heroes. I challenge them. I throw in something extra just to find the breaking point. And they don't break! They take it in stride. And, you know, when we hired them in early 2000, competition was tough to get people because we were still at a peak of activity in the Bay area. Now I think they're very happy to have a job. A lot of their friends took positions at other companies that went under. Everybody understands that our positions are really dependent on our production, on following through with Tom's vision here and making it happen in the lab. And that's what we dedicate ourselves to.

PR: Would you finish the description of the run? And then tell us the potential problems?

JC: Sure. There's a burst of activity right now in the disease areas. They have to transfer things to the DNA lab to configure pools and then create the pools in a wet lab. At that point, the DNA lab will get a little bit of a break, because they've made this concentrated parent plate and then it will be able to be reproduced through dilution for as many as 10,000 assays. The primer lab is essentially a crazy house right now. In addition to all the work I'm asking them to do to keep the regular studies moving, I'm asking them to assemble these whole-genome scan racks, because once we make these, we'll have enough primer in a diluted state to do fourteen studies. That is probably enough for a year,

year and a half, based on our schedule, maybe even two years. But quite candidly, we're going to start the study, but I don't think we're going to be able to run right through it uninterrupted. The study itself would take about twenty-nine workdays uninterrupted, 100 percent dedication to complete it. My guess is it's probably going to take at least double that, because we have ongoing projects. We want to keep on running parallel processes here, and hopefully everybody will be happy. The danger is that sometimes when you compromise too much, you end up making everybody unhappy.

PR: So the twenty-nine working days are spread over a ninety-day period?

JC: Roughly speaking. A lot of the timing is going to be dependent on our ability to get the assays formatted for the whole-genome scan. As I indicated earlier, we are running behind. We need to get into catch-up mode, and we've ordered another robotic liquid handler to help us do that and a $30K dispenser that will probably gain us 30 percent or 40 percent time. We were hoping to have got that two weeks ago, and now I'm hopeful we'll have it next week. But I think it kind of points out to you how we're really doing this on the fly.

In terms of the whole-genome scan, if we run into a problem, what we would revert back to is using the more expensive instrumentation at a slower rate. I don't anticipate a problem with this new machine. We've brought in three different vendors' machines, and we've done as many tests as we could on a loaner instrument for a day or two, and we've made our decision on which one to purchase, and I'm very confident that we'll have it up and running in a matter of days. We're really gearing up to convert the whole-genome scan from what I would like to call the first-generation, slower method to a smart, second-generation method of doing 96 assays on a plate, which would mean turning the whole-genome scan around in two days.

PR: Oh, come on!

JC: No, two days. Now, that's really the simplest case of a single stratification—all cases, all controls. For every additional stratification that we do, we would add essentially another two days. One of the issues for us is our ability to cross Ravens, and what we're finding is that we don't get uniform performance for some of our assays when we cross from one Raven to another. We now feel we have a handle on it. We feel now that we've come to an understanding why certain instruments are what we call nonperformers versus performers. For instance, on one instrument of the thirty-nine Ravens, Raven 130, 30 percent of the assays that we put on there in a given day fail. They fail because their growth curves fail to reach a set, arbitrary amplitude. We have another instrument, Raven 105, that has a 5 percent failure rate. So what is different about these two instruments? With the help of Applied Biosystems' technical staff, we tore these guys apart, and we still didn't have the answer. We're hoping that they're going to follow through on this, and we're pretty confident that they will.

PR: At the end of this run, what is the data going to show you?

JC: The data is going to show us perhaps nothing at all. I'm just presenting you with the worst-case scenario. That's very possible. Very possible. If we made a mistake in how we stratified, we mixed individuals such that we essentially masked out the association, what we would probably get is a lot of false-positive markers that wouldn't reproduce in the next sample population we do. More than likely, we are going to be buried in positive markers. We are going to have so many positive markers that it's going to take a long time [*laughs*] to analyze the data, but we're going to focus on the ones that the biology directs us towards, and we're going to attempt to replicate them as fast as possible, because our policy here is that we don't have an association until we replicate it.

PR: So, if the stratification is wrong, what that means is that the pooling needs to be redone?

JC: Well, that's one aspect of it. The power of pooling is that you can do a minimal number of tests, but it requires up-front stratification, and it requires, obviously, some knowledge, in terms of information to allow you to put your pools together. But once we make these pools, they're made. You can't unmake them! You've committed very valuable DNA to this process, and it's important to get it right. I'm not worried about that. I'm more worried that technically we're able to detect low-frequency alleles in pools of various numbers, and some of our pools may go well beyond the fifty individuals per pool that we had targeted. I'm more worried about that. We've done a lot. We've done plenty of studies up front to indicate that we will be okay, but I worry about things like that. This hasn't been done before. Many things could go wrong. Today is a perfect example: I've been looking at data that's wrong. There's something wrong! We made a mistake. We've been doing this for nearly eight months now, but yesterday we made an error somehow.

Many days I'll sit and I'll try to think: "Did something big happen today?" And, you know, ultimately, I always say, "No, not really." I view this kind of like a crawl process that we work in incremental steps, and I can't pinpoint a particular moment in time so far and say, "Wow! That was a turning point." Maybe John would be better for that. His vision is more wide-range than mine is. I tend to get pretty narrow in terms of trying to get this to work, but I'm always looking for that turning point where I can say, "No doubt—we're going to do this." I'll let you know when I hit that point! I haven't hit it yet. I just haven't.

PR: Do mistakes keep people creative in a way?

JC: Mistakes always keep people on their toes. A small amount of error is a good thing, believe it or not, because it prevents the catastrophic errors from occurring.

Chapter Four

ETHICAL
AND SOCIAL
CONSULTANCY

Social relations are made, not given; this truism carries even more weight when such relations are forged in relatively uncharted domains where new institutional practices are being formed, as is the case with Celera Diagnostics. Social relations in such institutions are freighted with distinctive challenges. It is not a question of starting totally anew but rather of examining existing forms—for example, of trust and confidence between researchers—dividing those forms into smaller units, evaluating the utility and worth of each unit, and then assembling those considered essential, along with new ones relevant to the situation, into relationships and institutions that must then be put to the test in practice. The following two chapters explore deliberations on how a company like Celera Diagnostics is initiating, shaping, and integrating a range of relationships between itself and the pharmaceutical sector, other biotech companies, government regulators, and academic researchers.

While the interviews in these chapters touch on a number of substantive issues—patenting policy, government regulations, publication strategies for scientific discoveries—we think it would be helpful to underline a few general themes that run through these deliberations and inform the subsequent policies and practices. High on any such list is the topic of trust, or more accurately, the place of trust in relations between sectors presumed to have different and competing interests. In order to set the stage, some preliminary distinctions are appropriate, in particular the helpful one drawn by German sociologist Niklas Luhmann between "familiarity," "confidence," and "trust."[1] Familiarity is a constituent dimension of many, but by no means all, aspects of social life. This claim means, among other things, that a good deal of time, both formal and informal, is spent building familiarity, getting to know one another, as relationships develop among scientists, or between scientists and lawyers, or between lawyers and patent officials, or between patent lawyers, et cetera. Achieving familiarity takes time, and by definition some of that time cannot be explicitly goal-directed activity; hence the frequent dinners, lunches, walks, and exchanges of all sorts that go on between people beginning and sustaining collaborations. This requirement helps to situate the incessant and costly travel that business people, lawyers, and scientists accept as a necessary part of their lives. Familiarity requires repetition and shared "down time," and, whether consciously or not, produces a degree of consistent and mutual expectation.

The achievement and successful maintenance of such taken-for-granted expectations result in confidence. Confidence, Luhmann observes, operates in situations of possible disruption or danger. Therefore it implies a degree of self-reflection and a decision to engage in forms of action and interaction that can be relied upon, because they arise out of achieved familiarity, but cannot be relied upon with certainty. When two parties agree to a meeting, they are mutually confident that both intend to participate. Of course, if one party does not show up, then an explanation is expected. In situations of confidence, Luhmann observes, it is "meaningful to reflect on pre-adaptive and protective measures,"[2] to attend to possible dangers, breakdowns, or disruptions.

Trust is something quite different than confidence, although the two are often closely connected in social interaction. "Trust," Luhmann writes, "depends not on danger but on risk. Risks emerge only as a component of decision and action. They do not exist by themselves."[3] It is precisely in the deliberations about what forms to give to collaborations that this balancing of trust and risk is vividly displayed. The decisions taken can never be entirely cast in terms amenable to a calculative rationality. The reason this zone of indeterminacy exists is that relations of trust are Janus-faced: they are embedded within relations of familiarity and confidence that were built over time, but the exact character of the attendant risk will only be known in the future, after it has materialized.

In this light it is exasperating to observe that if there is one claim that almost anyone reading the press (including the scientific press) or the more scholarly literature on the contemporary biosciences would accept without hesitation, it is that the bonds of trust and openness that used to characterize the relations of researchers have been corroded by the rise of commercial science. This claim has required little empirical or theoretical justification to be accepted, as it is apparently self-evident to many within the academy and among the general public, both in Europe and the United States. Indeed, a handful of surveys on the topic have provided a smattering of qualitative evidence to support the claim. Nevertheless, for two major reasons, we should pause and reflect a bit more before accepting it at face value.

The first reason is that, today, the claim that the sciences previously were open and trusting is rarely, if ever, questioned or scrutinized. Scholars have curiously forgotten (and the media never knew) that one of the founding discussions of the social studies of science turned precisely on this issue: Did the so-called universal norms of science identified by the sociologist Robert Merton actually characterize scientific practice? The debate was quickly resolved in the negative and has basically been forgotten, but its insights are noteworthy. The few empirical studies we have show that Merton's norms—"universalism," "communalism," "disinterestedness," "organized skepticism"—may well be values that scientists cherish, but their actual work is not guided exclusively by them: "Common access to information is not an unrestricted ideal in science; it is balanced by rules in favor of 'secrecy.' Intellectual detachment is often said to be important by scientists; but no more so than strong commitment. Rational reflection is seen as essential; but so are irrationality and free-ranging imagination. The use of impersonal criteria of adequacy is often advocated; but the necessity of personal judgments is also frequently defended."[4] It is worth underlining that the sociologist Michael Mulkay argued this point in 1980, before the advent of the biotechnology industry.

In the epilogue to his path-breaking book *A Social History of Truth*, sociologist historian Steven Shapin argues that there is an omnipresent nostalgia in modernity for the "world we have lost," and that nostalgia blinds us to the actual endurance of many practices. And this brings us to the second reason to be cautious about sweeping claims to perdition; Shapin sagely cautions that while traditional ties have been eroded in many sectors of social life, judgments of character and bonds of trust operate today in complex ways that we know little about. It is curious that there are almost no studies of how venture capitalists make decisions or socialize with each other; how the patent office works on a day-to-day basis; how policy, deliberation, and case evaluation are articulated within the FDA, drawing on relations of familiarity, confidence, and trust as well as on regulations and statutes. One thing, however, is certain: interpersonal connections, politics of all sorts, vocational commitments, ethical engagements, affective dispositions, and the like play an important role in how such institutions adjudicate claims of knowledge.

The most cursory acquaintance with the day-to-day operations of these diverse sectors reveals that how to draw, negotiate, and maintain a prudent proportionality between secrecy and openness, self-interest and impartiality, trust and mistrust is one of the main issues that those working in these sectors struggle with on an ongoing basis. Another thing is certain: we will not learn much about the practices of such institutions by relying on public opinion and the received belief of moralists.

Our aim is a modest one: to reintroduce some questions that are presumed to have been answered, and to make available aspects of one instance of how such matters are being addressed today.[5] After all, one of the main challenges facing firms like Celera Diagnostics is how to assemble and coordinate a diverse set of skills, assets, interests, dangers, risks, and hopes into a working apparatus. The following interviews provide a glimpse into this process of deliberation, assemblage, regulation, and production as well as the ties of familiarity, confidence, and trust that permeate them.

Interview with Paul Billings, Ph.D., M.D., March 21, 2003

Rabinow met Billings in 1991 when Billings interviewed for a position as advisor for ethical, legal, and social issues at Lawrence Berkeley Laboratory in Berkeley. Charles Cantor, who headed the Department of Energy Human Genome Initiative at the laboratory, invited Rabinow to attend Billings's job talk. Other factors intervened, and the position never actualized, but Billings maintained long-term professional contact with Cantor (now scientific director of Sequenom Corporation), and Billings and Rabinow became friends and engaged in innumerable discussions about science, technology, ethics, politics, culture, aesthetics, and sports over the course of the human genome project's natal decade.

Paul Billings received his M.D. and Ph.D. degrees from Harvard University in 1979. He is a founding fellow of the American College of Medical Genetics. He is a cofounder of GeneSage, Inc., a company that provides links between genomics research and health professionals and consumers. He has been a member of the faculties at Harvard Medical School, the University of California at San Francisco, Stanford University, and the University of California at Berkeley. He was a member of the NIH/FDA Task Force on Genetic Information and Insurance, has been a technical advisor to the National Association of Insurance Commissioners and the NIH/FDA Recombinant DNA Advisory Committee. In June 2003, Paul Billings was vice president for life sciences and clinical affairs for WIPRO HealthScience. He chairs the board of directors of the Council for Responsible Genetics.

Consulting

PR: Let's start with your relation to Celera Diagnostics. What do you do for them? How did it come about? And where do you think ethics consulting fits into the larger picture?

PB: It came about in part because you introduced me to Tom White, two years ago. Tom said that he would like me to consult but was vague about exactly what he wanted. It wasn't exactly ethics. I don't think there was anything about ethics in the consulting agreement. I agreed to consult on and review projects of their choosing for whatever purpose they chose to use me. My relationship has been centered on two things. One was to help them find tissue collections. I originally helped them with some Alzheimer's collections; I made suggestions about other samples that I knew about that might be available. The second thing that they've asked me to do is to review their research agreements with collections for what one might call ethical issues, but which I would call problems that could come back to bite them in the marketing or in later intellectual property discussions.

PR: So ethics in the sense of possible legal/social implications?

PB: Exactly. For example, they could be considering entering a collaboration to use a set of samples from somebody, and the question is, Did the people who collected the samples do it in a reasonable way, or is somebody going to get annoyed if they know that their samples are being used by Celera to find something else? I work primarily through their director of medical genetics, Linda McAllister. I have a retainer that they pay every year.

PR: So could you give us an example of your work?

PB: I'll tell you about the last project. They used my services with something called Genomics Collaborative, Inc., an outfit in Boston whose business plan is to collect samples. They go to various clinics and make deals with those clinics to get samples from patients. They standardize the information that they collect and then run a DNA sample of a certain kind; then they sell those to companies like Celera. That is their entire business model, and some people think it's a wonderful business model, but other people think it's a terrible business model. So, for example, if Celera wants to do a study on rheumatoid arthritis, they might ask Genomics Collaborative if they have samples. Are they the right samples? And have they consented these patients properly for this kind of work? Since you would think that, given that it is their business model to work with companies, they would have thought about everything, and Celera wouldn't need anybody like me to be on call. However, Linda McAllister says she doesn't think Celera can use certain samples because the collectors didn't follow HIPAA (the privacy protection for disease patients who are being treated in these clinics). So they asked me to review the documents for the collaboration and then to be on the call when they were discussing these issues with Collaborative Genomics. I told her, in this particular case, that Celera had nothing to worry about. The collectors had followed enough of the procedure for Celera to have plenty of cover for any adverse outcome, any disgruntled patients. There were certain little things they could do to make it look nicer, but fundamentally, they were fine.

PR: It is amazing to me that by 2003 there isn't a standardized form for such operations. There has been such an intense ethical and legal debate for years around these topics.

PB: The companies are oblivious to the debate. This debate to them is like angels dancing on a head of a pin. They don't even know about it. Their experience with IRB [institutional review boards] varies. The Berkeley IRB behaves entirely differently than the University of Mississippi IRB.

The take-home message from this is that, like in everything else, politics play a role in the IRB's work; at Berkeley the politics are of one arcane sort and at the University of Mississippi another. There are different kinds of power structures; if you really want to look at ethical decisions, you ought to look at the kind of political culture of where these decisions are made. In general, the companies want to do as little as possible.

A second example of work I've done for Celera has to do with a clotting study Celera wants to do. It turns out that in the Netherlands or Denmark—I can't remember—Finland—somebody has a huge system of clinics in which he's been studying the deficiency of Factor V, which is a clotting factor, and he's got 2,000 well-characterized patients, and he's been collaborating with other people. And they've published in prestigious journals. So Linda made contact with him, and he sends her the documents to start the collaborations. She looks at the consent document and it is, like, one line: "I consent to participate in the research," end of story. Now, that is inadequate by any standard. By the current ethos of consent, you're supposed to spell out what the risks and assets are. So the question becomes, Is this a shorthand for a document that they send along which is back there somewhere? Or do they explain it to patients without writing it down (which is less attractive)? Or is it simply paternalism that makes people sign the one-line consent form: "Oh, yes, you're the professor and we'll do anything you want!"? Linda says to me, "Can you help me figure this out?" I asked her if she wrote the collector and asked him, and she said, "Yes, but he's slow responding." So I suggested, "Well, why don't we try to find out from some people in Vermont who used the collection, went through the IRB, published something where they say that the IRB reviewed what they did? So why not ask what the IRB thought in Vermont?" It takes weeks to find the people in Vermont; it takes another few weeks to get them to respond, because they want to know what the hell you're asking about. If you don't ask about some esoteric finding of their paper, then they don't really want to talk to you, because they think you're going to cause trouble. IRBs are public, but to get the information from the proceedings of the IRB you need a Freedom of Information Act kind of thing. So that's another thing I did for them.

PR: Why did Celera Diagnostics hire you as a consultant to do this work when you're painting yourself as being well informed but not a lawyer?

If Celera's business and scientific strategy is so crucially based on tissue samples, why aren't they doing it themselves?

PB: I think that Tom drove this. He felt that I was sensible, that I was not an overreactive lawyer, that I was not a philosopher-ethicist who's going to wring my hands at everything. And it was also possible, by forming a relationship with me, that if, over time, there were other uses for what I call my translational role, they could tap into that. They'd have a contract in hand, and they could just change a few words and get me to do something else. So I'd be interested to know what Tom would say, but that was my fantasy about what they really had in mind.

I think that Celera was structured by the thought that they knew all the moves. They were Roche Diagnostics; they were just looping over, lock, stock, and barrel, to Celera Diagnostics, and what they hired was a basic research force. Linda McAllister, for example, has never done what she's doing now, which is clinical research. She was a molecular biologist. She was a gene discovery jock. She worked at Affymetrix, I think. She'd never done real clinical validation like finding genes and then trying to validate them as tests. I think that Celera recognized, maybe smartly, that I'm not a high-cost item for them; they could get an occasional validation of whether they've gotten off the track or not; whether in their zeal to do basic validation work and basic discovery work, they had stepped into a pile of shit—as some companies have, by the way, whether it's taking slave labor samples from China or whatever.

PR: Would you agree that the basic ethos of Celera Diagnostics is to prefer to do things legally and ethically?

PB: I think they have a clear view of the standards. Some other companies would like to tear down the standards. I think that Celera has decided to play the game by the rules. Do I think they respect the rules but wish they lived in a world that did not have rules? Yes, I think they'd like to live in a world that didn't have rules, but I think that because of their previous experience at Roche they know that it is possible to go through that process and get to a product. They send people to see the FDA on a regular basis.

PR: But you also think—I know, because we've talked about this many times—that these regulations are ineffective.

PB: Well, I think the fundamental question you asked was, Is their attitude about playing by the rules different than other companies? I would say it is informed by the fact that they have played by the rules and they know they can get through it. A lot of these companies have never been through the process and find each step daunting. There are several levels of complexity here. The creation of your product and the information that you use to create the product constitute one level of complexity, and another is all the manufacturing stuff. Now, the regulations are probably overdone; if you left it entirely to the market, a lot of these people would do exactly what they're doing anyway, because if you screw up [*laughs*], then you're out of business. You know, you kill a few people—you can say it's too bad you killed a few people, but the bottom line is the company goes out of business. Celera Diagnostics is one of the few companies that actually has gone through the processes. But they are also changing the target. Apparently, the CDC, through the Clinical Laboratory Improvement Act—CLIA—has propagated new rules for genetic tests that are very different than what was there before.

PR: Could you tell us about these changes?

PB: The new rules have made the stringency of the lab and the standard for good manufacturing practices in the labs much higher; the level of information, the content of the test, and of the counseling around the test has also been increased. The new rule says that if you do a test for cardiovascular, the company doing the test has to make available to the ordering physician an expert, not only in the test, but in the subdiscipline of cardiovascular genetics. And what does that mean? How many experts in cardiovascular genetics are there? There are very, very few.

Corporate Ethics

PR: Okay, now switch hats to Paul Billings, head of the Council on Responsible Genetics. Do you see any distinctive ethical issues at stake here?

PB: Why would a company spend money on somebody like me and bother to even think about these ethics? It's only marketing. It has to do with public relations and marketing. It's not because they think if you obey the rules you get a better product. I don't think they buy that.

PR: Do you buy that?

PB: I think it's possible. But you asked the question whether there was any value to compliance with regulations or ethical standards, and I'm implying that I don't think so particularly. They've done it, and their reason for compliance is to stay out of trouble.

PR: Then Celera Diagnostics is a straightforward company where reputation is a valuable asset in business. And that ethos, at least for this kind of business, makes good sense.

PB: We can look at it defensively, or we can look at it optimistically; some companies comply, and some of these rules do add value.

PR: So as an advisor to Celera, you would advise them to continue to do what they're doing?

PB: Listen, if you take the high road, you ought to beat the drum that you're taking the high road. That's basically how I feel about it.

PR: So either you are a pawn of capitalism yourself or Celera Diagnostics has what it says it has, a fairly well-thought-out, reasonable approach. Many of our dear friends on the left would accuse you of being part of this machinery. You are head of the Council on Responsible Genetics. You were one of the first people to identify the risk of job discrimination based on genetic profiles. So your credibility is high—but are you being used?

PB: Uh . . . [*Pause*] . . . no! Because the fact is, I mean, uh . . . uh . . . uh . . . first of all . . . um . . . I mean, if we're having a discussion about how health science or life science interfaces with healthcare delivery in the United States, let's have that discussion—in which case, I'm a critic of what we've done, because I think we need to fix the financing system. That is a separate issue in my view. I would say that if you take my concerns about the application of genetic tests for discriminatory purposes, my analysis has matured over the years: I would say that the preponderance of evidence now would suggest that problems that existed then and exist today are problems of complex systems. In a complex system, even if you want to make something as fair as you can, the system is going to continue to discriminate against the people it has discriminated against before.

PR: American society's disposition is to continue to be socioeconomically and racially unjust. The genetic engineering revolution, or whatever you want to call it, has contributed less to social change than one might have either hoped or feared.

PB: That's a good point. Let's take this issue of genetic discrimination. Why on earth isn't there more genetic discrimination? There could be a lot of answers to that question: one might suggest that good people thought about it and subverted some of it. I think much of what we saw originally was a complex system that was making some errors, and there's been some error control, and maybe the systems are a little better now and more informed. However, maybe there are just not enough tests out there, not a big enough industry yet for the issue to rise to the surface. People say that in those places where these technologies are getting more common, like in forensic uses, or in prenatal settings, you are beginning to see troubling trends out there. The ethical issues are coming out of the stratosphere and are hitting the ground. As an empiricist, I think you have to allow that as a possibility.

PR: I do! It is exactly my position. Yet, the U.S. Senate is passing legislation against genetic discrimination, legislation sponsored by Senator Orrin Hatch! Yikes! Dr. Billings, are you, or have you ever been, a part of the capitalist machinery of world domination, exploitation, and subjectivation?

PB: Look, I guess the way I would respond to that is [*pause*] . . . you know, we have lots of problems which we ought to be fixing, problems with our economic system and our healthcare system, and this affects not only disadvantaged people. I'm deeply concerned about the idea that we could do really good biological science, come up with something really wonderful for some disorder like diabetes, and that a third of the population would have no access to it or have very limited access to it, and that across the globe hundreds of millions and maybe billions of people would have no access to it, because of pricing and patenting. I think that while such inequities occur in any number of other spheres of corporate, commercial, social, and economic life, for it to happen in the biological and life sciences is really repugnant. But I see that as a structural problem of society, and I don't see that as uniquely a problem of the marketplace in the United States. This is a global problem. I do not find the market strategy of Celera to be unique or particularly rapacious. I think they have very high standards for a scientific enterprise that's trying to play in the real world.

Futures

PR: We are in agreement. What else would you want to add about Celera Diagnostics?

PB: One real question that remains is that since the founding of Celera, there has been another generation of biotech companies that are "epigenetic" biotech companies. There is a company called EpiGenome and ten other companies that focus on more than genes and SNPs. It has less to do with the discovery of a gene and the commercialization of a test from presumably the variation of that gene—which is fundamentally what Celera is about, though Celera would say also that they're about expression levels, and that then gets into the epigenetic business. Rather, there is a more complicated question, which is, Aren't we really after some interactional event in a person's life, in a set of environments that are mostly common but also have some developmental uniqueness? I think you could ask Celera, "Are you wrongly focusing on the fundamental unit, the gene, and making a strategic mistake by doing that because the action is somewhere more downstream?" Health is probably related to functional genomics; it is interactional, and their business model distorts that.

Further, Celera believes that those people who take care of patients—the doctors and laboratory people of the world and so forth—when presented with Celera's argument for their test, will say, "Yes! Whatever we're doing now to take care of people would be made much better if we had Celera's panel!"

PR: But it doesn't have to be much better. It only has to be competitively better. To succeed in a neoliberal private sector, you only need enough people who are capable and willing to pay. Some tests that could be done for $100 will be marketed at $2,500. Would you spend that money? Many people would.

PB: I'd push them on that particular issue, because, as you know, I am in favor of consumers having the choice to spend the money on tests that have interest or value to them. We're entering an age when informed consumers are a much larger player in medicine than they used to be when it comes to diagnostic testing. And there's any amount of evidence that supports that. They do not want to set up Celera Diagnostic kiosks in Whole Foods, where you can go in and touch the thing and you get your blood clotting test and your prostate test and whatever. They don't want to do that. Some companies do want to do that, but Celera wants to go the route of the traditional medical model, in which your physician orders your test because your physician believes that there's evidence that the test will be of some value to you. This line is contrary to the data which says the physician, eight times out of ten, gives you the test you ask for, whether he thinks it is of any value or not! Because he wants you out of the office as quickly as possible, and he wants you to be satisfied. He doesn't want you to call up his medical director and say, "This guy is a jerk, and he doesn't give me the test that I want."

PR: Whatever the management at Celera wants is to a significant degree irrelevant, because there's a time frame: in x number of months, there will be SNP indications for x, y, z number of risk conditions. Within a finite number of years, there will be more of these correlations, and the state of healthcare in the Western world will not depend on diligent scientists of this world; it will be the businessmen of the world who will have the most say on that.

PB: Sure, sure! But the issue of whether Celera Diagnostics will be in this business ten years from now will significantly depend on integration with unglamorous institutions like LabCorp. The questions that Celera Diagnostics is trying to answer for the system are the slightly more sophisticated questions. They are trying to specify their niche. Celera is looking to define their niche and make money in that niche rather than change the world. Celera Diagnostics and Genomic Health and a couple of other companies are now talking about addressing questions that are slightly more contextualized. For example, suppose you have prostate cancer. Now what kind of prostate cancer do you have? Should we give you this drug and not that one, because you're either a responder or a nonresponder to that drug? Or the relevant question with prostate cancer, which is, 80 percent of prostate cancer is benign, 20 percent is not benign. Who has the not-benign stuff, so we don't kill them with the treatment? What I'm suggesting to you is that Celera Diagnostics is a second-generation genome revolution company which has smartened up by scaling back its genetic fundamentalism, by emphasizing the potential genetic contribution to relevant questions of practical healthcare biology. Are we going to use this powerful set of biological methods and tools to help? I would say that the verdict is fundamentally still out. I don't think there's any argument that can be made that when Paul Billings gets ill in a hundred years, my concerns will be any different than they are today. What is my illness? What can

I do to make it better without killing myself in the process? The relevant questions for health and biology are going to be the same for the end user, I believe. The issue is the method of developing new answers to those questions.

The tools allow a certain set of questions to be asked, right? You have very fine molecular tools, so you can ask genetic questions. If your first-generation question revolves around the idea that genes matter, then you sequence the genes. But when you have sequenced the genome, you see that that hasn't changed the world and you figure out the next question: Do genes matter in the progression of a cancer or in the response to a drug? And you apply the methodology, you get to a set of answers, and the answers are adequate for some questions and inadequate for others. Then, hopefully, you start saying, "Well, let's figure out in what ways the question has been too simple, and let's ask the next question, which is, Are there environmental agents out there that we haven't identified?"

At the end of the day, Celera Diagnostics has a very significant advantage: the connections they have with Applied Biosystems, a hugely profitable business. They also have a relationship, however problematic and of questionable value, with a discovery shop, Celera Genomics. The health of Celera Diagnostics is tied to Applied Biosystems' health. If I were someone who was thinking about their investment in Celera Diagnostics, I'd be reading the quarterly reports of Applied Biosystems and trying to get a sense of whether they're going to bounce back. That instrument business went through this boom because all of these biotech companies were being founded. Today there are no biotech companies being founded, and people are not buying the second, third, and fourth generation of AB's instruments. That is going to have a very negative impact on Celera Diagnostics. You know that as soon as they've got a product out there, they'll get bought out.

Chapter Five

CONFIDENCE
AND TRUST

Celera Diagnostics, like all of its competitors, has a busy legal team. From negotiating and drafting contracts for the use of each sample set to submitting patent applications for the components of products still in the research phase, the company's legal team operates in many different domains. Like a disease group head, Victor Lee, who oversees the legal team, is both a strategist and a technician. Legal procedures can be approached a number of different ways. The terms on which Celera will sign contracts, the content and quantity of discovery that the company deems sufficient to justify a patent, the style of negotiations with patent officers, with other companies, or with research institutions are all subject to judgment. Victor Lee's goal, like that of many department heads at Celera Diagnostics, is to standardize strategy in order to contribute to something stable, consistent, and predictable. Thus Lee, in 2003, a year in which Celera Diagnostics is still being formed, is laying down the foundations for the company's future legal procedures.

Lee is involved in making sure that Celera Diagnostics is not vulnerable to lawsuits. At a recent meeting in a Celera Diagnostics conference room, a member of his department explained the ramifications of changes in the Health Insurance Portability and Accountability Act of 1996 (HIPAA) on CDx's research protocol. The law tightens restrictions on the transfer of information that could be used to identify patients. Although the main thrust of the new regulations has no direct bearing on CDx, the changes affect many institutions from which CDx receives samples. The presenter condensed hundreds of pages of legal constraints and mandates down to twenty simple slides that explained the important concepts and changes contained in the new law.

The meeting was casual. Most of the attendees exhibited familiarity and a sense of humor, as they ate their catered lunch, listened, and asked questions. Across the table from the presenter, who was standing as he explained the slides, sat Victor Lee. Not an expert on the details of this particular law, Lee was nevertheless clearly the one most capable of identifying the issues pertinent to CDx. He addressed his questions both to the disease group heads and to the presenter, in order to determine which current lab practices might be affected by the implementation of new standards. Those present at the meeting perceived the new regulations as ambiguous in relation to CDx's research. Although the law lays out a number of clear guidelines for defining the kind of information it is designed to protect, it also requires the exercise of case-specific judgment. After listing the defining characteristics of information that might potentially be used to reveal a patient's identity, the law states that it is not applicable when "there is no reason to believe that the information can be used to identify an individual." Thus, the law leaves room for judgment and contestation. Furthermore, without precedent cases on which to base legal judgment, affected institutions and organizations have to interpret the law for themselves.

Lee argued that CDx's researchers should play it safe. He questioned the disease heads about the information they receive that may identify individuals and asked whether that information was really necessary for the company's research. Most of the disease heads responded that they did not require any such information. A few, however, specified that they used the dates of hospital events to assess samples. Though it is doubtful that these dates would qualify as information that can be used to identify individuals, Lee recommended that Celera request institutions handling personal information to eliminate all dates on materials sent to Celera.

Interview with Victor Lee, March 7, 2003

PR: Why don't you tell us how you got to be in this office?
VL: I was in private practice for a number of years in New York and then in
Silicon Valley; I joined Roche Molecular Systems in the fall of '99 and

became their chief patent counsel. So for a couple of years I had worked with Tom. I did not work directly for Kathy. I reported to the general counsel, who reported to Kathy. I came to the office one Friday and heard that Kathy, Tom, and John had resigned. It was a complete surprise. I still remember; it was in November 2000. I arrived in my office on a Friday and one of the secretaries said, "Have you heard?" I said, "What?" "John and Tom and Kathy resigned." That was a pretty surprising thing to me. I had only been at Roche at that point for about maybe fifteen months, so I was still relatively new. But I knew how well respected they were and how well they had done there. I had no warning, because I was relatively new. It was not until the following week that it became clear that they had joined the Applera organization. And then life sort of went back to normal. It was not until, I guess, February 2001 that I saw an ad in the *San Francisco Chronicle*. It was fairly large, with a list of positions. There was one category they were recruiting for, "patent specialists"—an unusual term, because you would expect either a patent attorney or a patent agent, but a "patent specialist" doesn't really say a whole lot. It piqued my interest, and after some discussions with friends and colleagues, I decided that maybe I should put in an application. It was a little difficult to do, because you would think that at that stage of your career you don't really look for a job in the newspaper [*laughs*].

PR: Basically, they were constrained by legal considerations?

VL: I think they had an agreement with Roche when they left that they could not actively recruit, so they were very careful in how they handled that. Eventually I heard back from Kathy, and I came in for an interview in the spring. I got an offer in May and joined in July. I had a desire to follow this up, as this place was taking shape.

PR: But you felt then no dissatisfaction with Roche? Rather you saw an opportunity?

VL: Right. From a personal perspective, I was reasonably happy at Roche. I went from a very busy job in private practice to Roche as in-house counsel, and things were a lot more manageable to me. On the other hand, I think over the course of about a year and a half, I began to feel a little bored. You know, one can never find really a perfect job. Either you're too busy or it's not exciting enough. My usual way of approaching jobs is every job I take, I go into it thinking that I will retire there. I don't really think of using it to jump to some other place. So after about a year and a half, to think about changing jobs again was a little bit difficult for me. If it weren't for this group of people, I probably would not have left Roche at the time.

PR: In the interview, did Kathy outline in detail what she wanted?

VL: Well, she obviously wanted me to do the patent work. She also wanted me to take on more of a general legal responsibility for the organization. I was strictly a patent attorney at Roche. I was not even doing licensing agreements, because that was a different area that another attorney was handling. She wanted me to take on the licensing role as well, and to the extent that there are general legal issues, she basically wanted me to be that lawyer for the company.

PR: "General legal issues" means what? Not leasing the buildings?

VL: Well, in a sense, all of those things. Perhaps I wouldn't be actively negotiating the lease, but if there were a lease agreement, at least it would pass through my department so that somebody under my watch can take a look at it. My responsibilities included general labor law issues, employee issues, recruitment issues, basically anything having to do with legal liabilities. This position was not a general counsel position, because the way our organization is structured the legal department here is pretty much a technology-oriented legal department, i.e., IP—intellectual property, patents and licensing. There is an Applera corporate legal department in Connecticut that handles general law matters under the General Counsel. The same is true with Applied Biosystems. Their legal department is composed of patent attorneys as well, and general law matters—true labor law issues, corporate governance issues, investor relationships, SEC filings, 10Q, 10K, all of those matters—are handled out of Connecticut. Kathy wanted me to at least be a funnel for these things to pass through, which itself presented a challenge to me. I never really wanted to be a general lawyer, but I wanted the opportunity to learn something different and new to broaden my skills. Also, there was the licensing aspect, which is something that I always thought I wanted to learn but never really got around to; that was clearly on the table in this position. So I came, and it has all come true [*laughs*]. Fortunately, we have help from various sources in either outside law firms and also via the Connecticut legal department. So I'm really learning a lot about a variety of things that as a patent attorney normally you would not need to deal with.

Clinical Samples

VL: In the past year and a half we have spent a lot of time on negotiating and drafting agreements with various institutions to acquire tissue samples for all of the disease area groups. Before you patent anything, you've got to make discoveries; and before you discover anything, you've got to have the raw materials to work on. So from a very early stage, acquiring clinical samples was the bulk of the work that a variety of departments here were working towards. Business development was trying to identify institutions that had the right sample sets for us. The research and development people were using their contact with scientists to identify the right samples. And then I came into the picture, along with other lawyers helping me, to really finalize the contractual relationships with these institutions.

PR: And if I understand correctly, there's the usual IRB [institutional review board] and patent and technology transfer issues about how to have legal access to tissues.

VL: Right. I think the basic issues—policy and human subject protection issues—are in place; you certainly cannot simply take blood or tissues from people without informed consent. We look very carefully at the institutional board review approval to make sure that everything is in order. Although these samples would never be used as the direct commercial embodiments (because we're not in the business of selling tissues or blood), they will be used in discovering, well, in making inventions that will ultimately go into commercial products. So we have to be comfortable that the patients have granted consent that their samples could be used in research projects that could indirectly result in commercial products. If the informed consent is not to our satisfaction in that regard, we generally don't want to touch it.

PR: Is that form relatively standardized now?

VL: Well, it really depends. Most of the informed consent agreements are pretty good; we are pretty comfortable with them. You want them to say, "I, patient, understand that the samples may be used in research that results in a commercial product." With statements like that, you're okay. But there are times when you don't have that statement. What you do have does not expressly prohibit the use of the samples for commercial purposes; you have to wonder what was said to the patient. With those it's going to be a case-by-case analysis. Recently, we've come across situations, particularly with institutions in Europe, where they use oral informed consent [*laughs*]—well, how can you know what was said and what was agreed to? So these are issues that we still struggle with, but by and large, when you're talking about the major research institutions in America (we've done a lot of deals with UCSF, Stanford, Yale), I think those cases are clear.

PR: Okay, tissue or sample collection is obviously the key for getting this project going in a rapid fashion. Has it gone well?

VL: Yes! Although I knew that there would be some of this work when I joined the company, I didn't expect to spend more or less 80 percent to 90 percent of my time in the first year on such agreements. In the course of that, I learned a lot; we must have concluded fifteen to twenty different such relationships for at least six, seven different types of disease samples. These are diverse disease areas, all of which will be used this year for our studies.

Now, how good a job have I done probably cannot be measured just by how many deals I've done. It's going to be tested in time [*laughs*] by these agreements as to whether or not I have sufficiently protected the company. It's something that I always worry about, because these things don't come to light, probably, for five years.

PR: Is that the time horizon?

VL: Well, I wouldn't say that there is a particular horizon, but in this commercial world, the fact of the matter is, it's not until there is money involved that there would be people coming out of the woodwork to contest things. So, when we have a successful product out there, three years from now, five years from now, that will be the time to see whether or not somebody will be challenging it.

PR: Would these be money challenges or patent challenges?

VL: I don't think it would be patent challenges. For example, who will get to own these discoveries? It is undisputed that the samples were collected by, let's say, UC San Francisco and they were the ones who provided the samples to us. Now, these are not necessarily fee transactions, although there's usually money involved to cover their collection cost—the nurse, shipping, and all of that; it can all get to be fairly expensive. But once the samples arrive here, that's not the end of the story. That's when the work actually begins. We have to decide on which genes to examine, what reagents to put together to ask the right questions to make the discoveries and the inventions. It is our position that since we are the ones doing the work, the discoveries are made by us. We should be in a position to be the owner of the results and the subsequent rights in such results. This point of view, however, would not necessarily prevent people from arguing that "but for my samples, you would not have made the discovery." Now, I've spoken to a number of lawyers about this issue. It's a relatively new issue, and it is an argument that people could advance, but we feel that this "but for" argument is not really that strong, because there are other samples of that type elsewhere. We just happen to have negotiated and contracted with one particular party for them, but the real basic discovery is made downstream. Generally speaking the way we address the IP is that whoever made an inventive contribution will be the owner of that contribution. You can never really write an agreement addressing every point. Nevertheless, we feel that since we're doing the work, we're selecting the genes, we're putting reagents to test those genes, we are positioned to actually make the discoveries, and we're positioned to be the owner of the IP that results from all that work.

PR: There are two issues that interest me: First, how are you approaching patenting or protecting multigene constellations? Second, how do you negotiate, protect, and orchestrate the relations among and between multiple groups working with you as concerns trust, confidentiality, and protection?

VL: The Alzheimer's working group is the first such relationship we entered into. And back a year and a half ago, when we first started to negotiate for these samples, we had less leverage than we do today because we needed material to work with, and so we had to open ourselves up to the terms of the other parties. I don't know whether I would necessarily want to be in a position where you have so many outside collaborators with every project. Not that we don't trust them, it's just that it's hard to control the academic community, where they are mandated to have the free flow of knowledge, as opposed to what we do in a commercial setting, in which you have to guard things very carefully. I'm happy to know that in the Alzheimer's situation people are working well together. The common goal is to make the right discoveries that are central to us as the commercial entity as well as to the academic scientist. So to the extent that we're all achieving that common goal, everybody has something to win here, and I think that's the case right now.

PR: It came up a few times at the meeting that a leak could be fatal for all the parties, but I take it that this is a situation that won't always apply or won't apply even with the same group indefinitely?

VL: Well, with the same group, I don't really know. Beyond a certain point, it's kind of hard to pull them back. I think we should just continue on the course, because we need to learn from all of these people. But it's not necessarily the way we're working with other diseases. There are areas where we involve more outsiders, areas where our control is tighter. Things are still unfolding.

PR: It is not ideal to have too many collaborators in the public realm because you have less control over things?

VL: Right. For me—and now I don't speak for the company as a lawyer—I prefer to take a more conservative approach, and if it was entirely up to me, I'd say, "Never publish anything! Never talk to anybody outside until we get the work done!" But I also understand that there are other ways of viewing things. In any case, everything is moving so rapidly these days that even three months of lead time give you a competitive advantage.

PR: I think this is something that many people don't understand. The argument is not that you'll never publish. These are not trade secrets. It is a question of one amount of months or another amount of months.

VL: Right. I look at the whole IP protection and patenting area perhaps differently from the public at large. I look at it as a way of advancing scientific progress. The reason being that you can trace the patent law back to the Constitution. Constitutional language states that the patent law was mandated to promote the progress of science. And the way it worked for the forefathers who put the Constitution together should be the way it works today: people who make these inventions are rewarded with a certain amount of protection, but the protection is of a limited duration. In the case of a patent, it is for twenty years. When it expires, it opens up the whole field for everybody to come in. It gives incentive to people who do the work, but ultimately it benefits the public. It costs a tremendous amount of money to do what we do, and we're hoping not only to run a successful business but also that our products will actually benefit human health. For us to continue to do that we have to be able to generate the revenues that feed back into funding additional programs. We feel we need to have the protection in place so that people cannot just take what we have labored on for so long and put out a product that puts us out of business.

Generally speaking, for patent protection you really don't need more than a few months to get it in place. Once you have that, you can certainly publish it without harming your chances. You can argue, however, that if you never publish, or if you substantially delay the publication, while it does not really benefit your patent position anymore, it at least keeps your competitors in the dark. It gives you more lead time. We, as an organization, are pretty open to the world, and we don't keep things secret forever—only to the point where it puts us in a position where we could not be easily gotten around, for lack of a better term.

PR: But publication also makes something public, and hence your competitors can't claim originality. That's another strategy, right? Put things into the public domain, and then no one else can patent them?

VL: Right, exactly. That is another way of hurting others' chances of patenting, but the flip side of your argument is, when you put something out in public, it presents people with an opportunity to design around it. And that's why nobody can have a perfect patent, because there will always be improvements that make things better that would not fall within a patent.

Multigene Patenting

PR: Can you give me an overview of multigene patenting strategies?

VL: Our concept here is the "constellation." We envision that a product may require the detection of a number of markers. We don't know how many. It may require a half-dozen, it may require twenty, and it may require a hundred—though I don't suspect it will end up being that many. The picture will become clearer in six months and much clearer in a year. But let's say you need ten markers in a product, and these markers provide higher predictive value than any one or few of them in combination. You want to be able to protect that concept as intellectual property. The fewer markers you protect, the broader the patent is. If there is one critical marker of that group and it's the core, and nobody can make a product without that marker, and if you have that protected, in a sense that is the only protection you need, since nobody can make a product without infringing your patent. As you discover the additional marker that add more value to the final constellation, it is also important to cover them, because the constellation will be your ultimate commercial product. You want to have a patent covering exactly what it looks like. If you have a patent covering the twelve markers in your product, then someone making a product with only eleven markers might not be infringing. That multigene patent is also valuable because if no one has ever come up with that unique collection of markers, it becomes readily patentable. But it is a little bit narrower, because all twelve markers have to be in a product to infringe your patent. I think we need both; we need to cover the individual stars in a constellation as much as we can. It is getting increasingly more difficult to do this, because the individual stars have been studied by a lot of people. There's a lot of information out there. So you need to have a collection of stars but perhaps not the entire collection, maybe a unique combination of three or four of them that people have not reported but you have found to have a higher predictive value.

PR: Reported together?

VL: Yes, that kind of patent is extremely valuable to us if we cannot protect the individual stars. Much more so than one that actually covers all of the stars in your constellation, because there may be stars in the constellation that are substitutable. So the first four are critical, but the other eight may not be.

PR: Is this approach established yet in terms of U.S. patent law? Are we moving from a single-gene paradigm to a multigene one?

VL: By and large, it's a fairly well-established area of the law. What is going to become fuzzy in this process is, if there have been reports about a couple of the twelve markers being useful and then a few of the other individual ones having been reported. Under the patent law, one can obtain a patent if it's never been done before in exactly the same way. But if what you are trying to patent presents an *obvious* modification, you also will not be able to patent it because of this obviousness argument. And obviousness is what patent attorneys spend most of their time fighting over, because it is such a nebulous concept. What is something that is obvious? How much more predictive value do you need for that constellation to be patentable over the individual markers that people had already hinted at?

PR: And all of these university groups, they may have their own marker that they've done a lot of work on. They may even know some of the functional genomics around it. But it's only a piece of something larger?

VL: Exactly. When you have quite a bit of information about each piece of the puzzle and you put the puzzle together, how much clearer does that picture have to be—when you have the whole picture together over the individual pieces—in order for you to patent that clear picture?

PR: Is this radically different for diagnostics and therapeutics?

VL: I wouldn't say so. I think this concept basically translates across all patent law, be it biotech, be it pharmaceuticals, be it vacuum cleaners, paint compositions. I think it is the same idea.

PR: So if you establish a successful patent, when you move from the diagnostics to therapeutic molecules you'd be well down the path to patent protection there as well?

VL: Right. One thing that I should mention is, in cases where the individual stars may have been reported, so that you can no longer patent them, or they have been patented by different companies or universities, then this so-called constellation patent becomes that much more important. Important, because you can overcome the obviousness hurdle if you show that when you put the stars together, the predictor value is much higher—if each individual star gives you a predictability of 10 percent of the people having that star developing Alzheimer's, whereas if you put them altogether 50 percent, that's a pretty large difference that you should be able to patent. Then, from a commercial perspective, it puts us in a position where we could negotiate better with people, because if I may need to license a patent from different companies for each of the individual stars in order for me to put out a commercial product, they similarly cannot put out a product without infringing my patent that covers the collection of these stars. So it puts you in a cross-licensing situation where you need each other, and that's where we can create value for ourselves: the patenting of this unique combination would put us almost on equal par with people who have patented the individual stars. If they really want to do something, they need our patent, and if we want something, we need their patent, so let's come together and talk and maybe we allow each other to operate and just compete out there by having a better product and not using patents to hit each other over the head.

PR: Is there competition from the commercial world for lining up university researchers to be collaborators because they're likely to be less ferocious competitors, or they're not likely themselves to try to put together a constellation or a commercial product?

VL: I have not looked at it that way. The way I've looked at it is, a lot of the smart people are in academia. These are the people that we work with as consultants and collaborators. We work with a lot of academic university scientists more because we need their brains as opposed to aligning ourselves with them so that they could not become our competitors. We think about other companies as competitors. And we deal with them differently from academics.

PR: Can confidentiality agreements be policed in any effective way?

VL: Well, it can only be policed if you can trace it and understand who is out there who knows what we're doing and there's no way that person could have known what they know without some leakage. How effective is this? We don't have the time to really go around policing. So there is a sort of a good-faith honor code that if we're going to work together, we're going to work together for each other's benefit.

PR: That is the part of it that's very intriguing to me, because there's an image in the larger world of profit-driven companies on the one hand and the knowledge-driven academy on the other. Having spent my life in the academy, I know it is not a place that's uniquely filled with honorable people. We have a situation here where there is a very interesting mix in which mutual trust seems to play an important role. There are patents, there is potential legal remedy, but the trust dimension seems still rather central for this to work. This is a surprise to me.

VL: There are a couple of things I should say. Not enough time has lapsed for us to see the whole process through yet, so we'll have to see what actually ends up happening. I hope everybody has something to gain out of this, but we've only been doing this for about a year and a half. What we bring to the table, what a lot of our collaborators see as unique, is that they couldn't possibly have done this by themselves or with anybody else. For academic scientists, they see the unique opportunity of putting themselves on a major publication. The driving force in almost all of these tissue transfer agreements is that they want to put their name on a major publication with us. When we come out with a publication in *Science* or *Nature* that the following five markers have been shown with such statistical significance with a disease, they want to be listed on there. That prospect becomes a driving force for them to abide by the rules, the ethical obligations, and the good-faith dealings with us.

PR: But you don't know after that?

VL: I have no reason to suspect that anybody would misbehave, but it's still a little early.

PR: So there is an initial convergence of interests, but you don't know yet whether once the publication is out and you turn it into a test, the desire to accumulate academic credit and material capital will arise.

VL: Right. I'm not saying that this will happen, but once we have made the discovery, we've made the publication, they may be waving their hands and saying, "What is my share in the profits?" We try to negotiate the agreements to prevent any of that, but every so often you do see a couple of people who [*laughs*] go after the personal gain in addition to what they went into the whole relationship for.

Patenting Strategies

VL: There are two aspects to the IP issues. One relates to how we protect the kind of discoveries and inventions we make here. The second concerns what kind of freedom we need from third-party patents as we configure our products for commercialization and what licenses we need before we can even take that step. So, starting with how we protect our own assets. What we are faced with in our discovery program is that we're in the business, at this point, of discovering genetic variations between individuals and then taking that a step further to determine what correlation these variations may have with specific diseases. There are hurdles in terms of patent protections with the first step. As you discover genetic variations between individuals, we must ask, "What possible utilities can you attach to these variations?"

Let me step back. The discoveries here are more or less done based on this so-called resequencing project that Celera Genomics, just concluded about two months ago. It was a jointly funded program between the three business units within the Applera organization. The goal of that project was to discover genetic variations from thirty-nine human individuals and a chimp because the original genome project, completed about three years ago, came from a very limited number of individuals' DNA. With the limited sources of DNA, while you're able to put together a series of genes that constitute the whole genome, what you don't see are the individual variations of these genes from individual to individual, from groups of people to groups of people. So we undertook this project, about two years ago now, to try to resequence all of the genes from thirty-nine individuals so that we could actually find the fine differences between people who all have the same genes, more or less, but they have very minute differences that may be the cause or link to susceptibility to certain diseases. There are a large number of diseases that have fairly clear evidence of a genetic basis, and that may be accounted for by such differences. So that first step serves as the basis for the research and discovery that we are doing here in Alameda. The process involves screening individual genetic differences against two sets of populations; one being individuals with a disease and the other a control set that does not have the disease. You look to see whether, when you compare all of these differences between the two groups of people, there are certain genetic variations that are more prominent in one group versus the other, thereby lending themselves as potential markers for predicting a genetic predisposition to developing a disease.

Now, how do you go about patenting all of these various aspects? How can you protect these genes that have minute variations in them? This is an issue that relates to patent law, which requires that each invention you file a patent application for must have a practical utility. That term, "practical utility," is loosely defined, and like most legal terms it requires various case law interpretations as to the precise meaning. But the way it is being viewed by the patent community right now is that a patentable invention has to have some utility that is of a practical nature in the real world, so that it is not enough simply to say that to have this genetic variation in this gene is useful for us to take it

further for additional research. But if you say that this genetic variation is used for predicting the disease, if you have evidence for that, or
that this variation is the cause of the disease, then that very well would
satisfy the patentable utility.

So when we start filing patent applications on these genes with specific variations, the first thing we have to consider is, What is the key
utility we can claim for such an invention? If it's a gene of unknown
function, it's going to be rather difficult. And if it's a gene of known
function, but you don't know what the variation does to it, you also
may face some difficulty. Is the normal gene function affected by this
mutation? Or is this simply a silent mutation that does not really result
in a functional change? So what it means is that even after you have
the genetic variation data, you often still need to know more than just
raw sequence information to be able to adequately address the utility
requirement that is imposed on every patent filing. And this has been
particularly difficult in the past few years, because the patent office has
raised the standards substantially higher than what people are used to.
So you are really asked to come forth with stronger information and
data to support what you are claiming as the utility for each particular
invention.

PR: That seems like a reasonable position on the government's part.

VL: It's a two-edged sword. It depends on the nature of the discovery you're
trying to protect. Some would argue that the Patent Office's standard
with respect to utility is higher than what the law normally requires,
because the utility requirement has traditionally been viewed as a very
low threshold. That is to say, very few things in life have no utility at all.
It's just a matter of whether that use is deemed to be adequate to satisfy
what the law requires to be the level of utility for something to be
patented. And that's where the dispute lies. Is use adequate as long as
you can assert it and define it? Or does it actually have to be in the
realm of something that is more practically relevant, from which you
can make a product that can directly benefit mankind and society in
some manner? That is a dispute I won't get into. But the fact of the
matter is, if you just submit a gene with the sequence information in it
and the particular mutation or genetic variation that you have found
to be novel, and yet there's nothing else in the patent application to say
what you could use that for, you're likely to be rejected—and rejected

repeatedly—by the Patent Office, until you come up with something that will be credible as having practical utility.

When we first tried, about a year ago, to look down the path to when these SNPs or variations would be discovered, we had to ask ourselves, What would be our strategy for protecting the inventions? Do we have to wait until we collect all of that data, conduct the disease association studies, find the markers that are actually associated with the disease, before we can file a patent application? That invariably would involve at least several months, if not a year. And, of course, everything is moving so quickly in this field, and there are always people out there working on the same area that may discover the same variations before you. They may not be working on the same large scale that we're able to, but if you have two hundred laboratories in various institutions, where people are working on a very small, defined area, collectively, they could compete with the large-scale discovery we're doing. So what we decided to do was to take advantage of the fact that we knew we would be doing a number of disease association studies, and we knew, for those studies, that we had our favorite genes for each disease that we wanted to study. The strategy that we put together was to start filing on these novel variations as they were discovered if they fell within our candidate gene list and assert the association of those particular variations with that disease as the utility for such markers right away.

Now, of course, as you do that, if you have a thousand genes for a particular disease that you want to go after as your candidate genes, and each gene may have three variations of interest, that's already 3,000 SNPs. We know that not all 3,000 SNPs are associated with the disease. We don't know what the number will end up being, but it could very well be less than 30—perhaps less than 1 percent. But we think that while we narrow down 3,000 genes or 3,000 variations to the number that we ultimately find support for experimentally, we can present a case in the applications as to how you start going through the process. While tedious, filing gives enough teaching and guidance for somebody to get there. It takes time and energy and money and work to get there, but getting there no longer requires an inventive contribution.

PR: So you're building relationships and networks with the Patent Office. There is a kind of instructional activity on your part to inform them as to where you're going. And even if they turn down a range of things, you're in dialogue with them?

VL: Yes, the strategy is very aggressive: going after a large number of these variations as they are being discovered, even before we do the experimental work. But the plan is always to follow up with the data, because we know these studies are being done, and that within a month, within a year, we will have narrowed down from the 3,000 to however many we end up ultimately using for our products. During the course of that year, the patent law provides a particular way of filing patent applications, as a result of a law change in 1995. Now you can file what is called a "Provisional Application," which is not examined. But within twelve months of the date on which the provisional patent was filed, you have to file a so-called regular application that is examined by the Patent Office. And if you don't file that within twelve months, then the first case is automatically abandoned. So we file the provisional cases with a lot of information in there, knowing that within twelve months' time we will have the experimental data to put in and eliminate the variations that do not end up being useful for that disease. Then you have a substantially tighter case, with data supporting what you're claiming, which is much more manageable in terms of examination and review.

PR: Is it ever public? Can anyone look at the applications?

VL: Not until eighteen months after the original filing. So six months after you've filed the regular application is when the patent is made public. So in essence, we're taking advantage of the patent system right now to try to lock us in with the early filings but then buying ourselves that twelve-month period to come back and really address the key issues.

PR: How much does it cost to file one of these provisional applications?

VL: It's actually very cheap. I don't have the number precisely, and it changes from year to year, but I think it's in the neighborhood of $150. I suspect that in the course of doing this, we probably will end up filing somewhere in the neighborhood of a hundred provisional applications. We'll probably end up with about ten or twelve regular applications.

PR: And how much does that cost?

VL: When you do the regular filing, it costs much more. It's in the neighborhood of $1,000.

PR: How much time goes into a provisional application? Are they pretty much boilerplate?

VL: Yes. We're using something that's fairly standard for all of them. We've been filing cases since last fall on more or less a monthly basis. This year, we may very well be filing on a weekly basis.

PR: And have you passed from the provisional to the regular?

VL: No. It's going to be quite some time before we get there, and when we get to that point, it's probably going to be even a year before we get a correspondence from the Patent Office telling us how they like it. So we are easily a year and a half to two years away from getting an indication from the Patent Office as to the patentability of these inventions. The reason why we try to do this massive filing is, if you get an early filing date, that's always to your advantage, for two reasons. One, any publications by us, or others, cannot be used to harm us anymore if they were published after our filing date. Whatever is considered to be potential damage gets cut off by the time you file. That is important, because it prevents other people from attacking the lack of novelty of your inventions. Second, as you're also competing with other companies who may be working in a similar area, if they should file on the same thing, you would be given potentially the priority status.

PR: And even if the competitor's is similar or even better, you're just in an all-around better position because of the priority issue?

VL: Right. That is why time is always of the essence, and an early filing in the patent area is always better than later filing. And that's why we can't afford to wait for the studies to be done. On the other hand, we're very comfortable that with respect to this utility requirement, once we have found markers associated with a disease, that utility issue is no longer going to be an issue, because the association will be viewed as more than adequate support for utility if you say that this marker can be used to predict somebody having a predisposition to developing a disease. There's no doubt in my mind that the Patent Office will accept that. Now, how is it going to work out? Time will tell. I mean, this is sort of new territory, and it probably will be two to three years before we can fully realize the effect of this strategy.

PR: But once you establish this framework of flow of preliminary and then formal applications, you could then have a large stream of things. And the examiners are likely to then be anticipating issuing patents: "Well, we gave them ninety-seven already."

VL: Right. So, it will be very interesting to see how quickly these applications move through the Patent Office, and I hope that two to three years from now, we'll be having a lot of patents issued to us.

PR: Could you just briefly explain the international dimensions of patenting?

VL: Obviously, patent protection in the U.S. is not adequate. You want to have worldwide protection—but how worldwide? In a lot of large drug companies, for a very important drug, they go worldwide. They seek protection not only in North America, Europe, and Japan but they go to countries like India and the Philippines and Argentina. For our businesses, we are not likely to do that. I don't think that the cost is justified by the type of products that we will have based on these discoveries, but certainly we want to have protection in Europe and Japan. There may be a few other countries that we also would do, such as Canada, our neighbor, Australia, which is a very industrialized country, but we won't go beyond a core of about half a dozen countries. The way you do patent filings for foreign countries is that all of these countries are members of a treaty called "Patent Cooperative Treaty," meaning that they have agreed to a common process of how patent filings can be done, based on the original filing in the country of origin. Once you have filed in the United States, within twelve months you can file a single foreign application.

PR: A provisional application?

VL: Well, the twelve months starts from the original filing. It is called PCT—Patent Cooperative Treaty—filing, where you can elect what countries you want this case to go into for issuance. So you will check off the box for Japan, Europe, and whichever other countries, and that case then has to go through some formalities at a particular office that looks at it to make sure that all of the pages are there, the drawings are legible; and it takes about a year for that to all be sorted out. And then it goes into the individual patent offices of the countries that you desire the case to be examined by, and you then deal directly with those foreign patent offices, be it the Japanese Patent Office or the European Patent Office, and then, hopefully, you get them to grant you a patent in their respective countries.

PR: Does this work well?

VL: It works pretty well. It's very much a part of the high-tech business nowadays. Everybody does it.

PR: And if you win here, then you're home free in Europe as well?

VL: Well, it's not always the case. While patent laws in many ways are very similar in terms of the requirements for patentability from country to country, each country does have its own peculiarities, so you cannot expect patents to issue on the same subject matter from all patent offices. And even if they do, the scope of protection may also vary. There are policy and ethical considerations. In Japan, traditionally, they grant very narrow patents. And that's consistent with the way the Japanese culture works, because the Japanese are very much in the business of taking what people have created but improving and making it much better. So their patent system allows coverage of subject matter that's in a much more narrow scope, thereby allowing other people to be able to take advantage of skills and creativities to build upon them.

PR: Do you go to Washington and have discussions with examiners?

VL: Often. After you file a patent application you wait for some time for them to respond, perhaps a year. Ninety-nine percent of the time they reject it for a variety of reasons: they will cite chapter and verse about why you have not met certain requirements—utility is one of them, novelty is another, things like that. Then you are in a position to respond by advancing arguments as to why they're wrong. Or, also, you can make amendment to your claims so that you could avoid certain rejections. We often set up an appointment to go into the Patent Office in Washington and sit down with the examiner. You have to understand that what they see is a hundred-page document in front of them. They don't have any of your background and your thinking, and they're just sitting there reading it cold. They are given a small amount of time to examine each case. While I can spend two weeks preparing the case, they are given somewhere in the neighborhood of two days to dispose of a case from beginning to end—from the first time they see it to the time they issue the case. A lot of times, they just flip through certain relevant pages but they won't really catch everything. So the best thing to do is to go in and show them the critical aspects of what you're trying to cover, and I've always found it to be extremely useful. They appreciate it. They can learn in a half-hour what they cannot learn in reading for five hours, and in the course of that you build up a relationship with them. You can often bring a scientist with you who can explain it to them. Most of the examiners are Ph.D. scientists themselves, so they like the interaction, they understand the scientific conversations, and you can often get much farther ahead with that one hour.

PR: So you're not emotionally damaged, then, when the mail comes in?

VL: Patent attorneys take rejections very well. We expect rejections. I've been doing this for about twelve years. There's only one case in my entire career where the first correspondence was an allowance.

PR: Were you shocked?

VL: I was shocked! And if you would talk to probably ten attorneys, ten of them would tell you that it's only happened to them less than 5 percent of the time.

PR: Hence, this personal interchange is not seen as manipulative but is welcomed? It is friendly.

VL: Yes. It is usually friendly. There are some applicants who go in and pound the table and try to accuse the Patent Office examiners of not doing a good job. I'm not of that nature. You don't always convince them, but if you are ever going to convince them, you stand a better chance doing it orally, because something is always missing in writing. There are times when what you need to convince them of is inherent in your document, but they just don't readily locate it. You need to just point that out to them, but you can also supplement with, perhaps, additional data that have come out since the filing that are consistent, and often that's compelling, because if they've seen it once, they may say, "Well, maybe that is your best experiment. How reproducible is it?" But when you come forward and say, "Well, no, no, no! This is not just a research finding. As you can see in table 1 on page 11, this is a product that our customer is right now providing as a service to laboratories in the country, and here are the thirty-five hospitals that send samples to them for testing for this disease." That is the kind of information that you don't always get to put in the application when you file but is certainly supportive of it. After a year or a year and a half of the case sitting in the Patent Office, data have been generated that you can bring to their attention. And that's the kind of stuff they're looking for.

PR: Do Patent Office people tend to stay there long enough for you to develop long-term relationships with them?

VL: Yes and no. There's a fairly large amount of turnover there.

PR: Because they're poorly paid?

VL: Well, a lot of them do go to law school at night to become attorneys, so compared to law firm salaries, yes, they are poorly paid. They are overworked at times, because there are so many filings and there's a budget constraint as to how many Patent Office examiners they can have. However, there are some people that I've worked with who have remained at the Patent Office for a long time. Perhaps they like the patent work. Also, the examiners are grouped by their expertise. So depending on what area of technology you are working in, it's usually a particular group of people who tend to examine your cases.

PR: Is this organized by disease or by technology?

VL: By technology. In private practice, in a law firm, I had a much wider group of examiners that I interacted with, because you are representing clients that work in different areas. But when you're working in a company where your inventions are more similar in nature, you tend to work with a smaller group, and that's good, because you really do get to know them and if you establish good rapport and credibility with them, then it helps you.

PR: Are the patent applications published online?

VL: Yes. It's searchable online in the Patent Office, and you can obtain hard copies, too.

PR: Do you learn anything from those?

VL: Yes, you do. Everybody monitors everybody else. You certainly learn a lot about what your competitors are doing and what stage they are at in terms of their patent portfolio and that kind of stuff.

PR: What percentage of your time is devoted to that?

VL: Not that much right now, because the focus right now is to get our stuff done and not worry about other people. But the time will come when we will have to start pushing these products out the door, and then that gets to the second aspect of what I started out with, that after you've protected what you need to protect as your own assets and discoveries, then, as you turn that into products for sales and commercialization, you need to know the patent landscape out there that may affect your ability to commercialize a product. Nowadays everybody files patent applications. There are all kinds of gene cases and methods cases, and so you never really ship a product out without knowing a little bit about who you need to deal with. Somewhere along the way, after we have spotted certain SNPs and markers to be of high enough value for us to really pursue products based on them, somebody will knock on my door and say, "Can we do some searching now to determine what parties may have rights pertaining to these markers?" Now, hopefully, these markers are the ones that we have discovered in our resequencing of the human genomes for the thirty-nine individuals so that we already have some strong patent position on them. That is not to say that there are no other potential dominating patents out there that we may need to license before we have the freedom to operate. So, at that point, we'll conduct some searching for patents in the U.S. first and then also in Europe—less so in Japan, because you don't really get translations so you have to pay somebody to do it. So we just do the initial searching in English-language databases and determine which patents of relevance are out there, who the parties are that own them, how strong these patents are, what the effects are, and if we should move forward with a product with a particular SNP marker if we believe that a particular patent could present a problem. If there is a patent that presents a problem, then we need to start initiating some contact with the owners of that patent to determine whether they would grant us a license, so that when we actually commercialize a product, we feel that we have all of the necessary licenses from owners of patents, so that we don't invite a lawsuit once we launch the product.

PR: Okay. When you start making alliances with companies like Abbott, do you start immediately working with their attorneys at that point, or has all of that work been done?

VL: Well, because of the alliance, we work so closely together with Abbott that our scientists work with their scientists on a daily basis. Our business development people work with them. I also have been working with the Abbott attorneys ever since we concluded the arrangement last summer, and on a weekly basis we have a scheduled conference call to talk about issues of this type. When we make discoveries of markers, we decide who should be doing some searching among ourselves and who is better positioned to approach that third party. Maybe Abbott has had a previous relationship, for instance. So we've been talking about these things all along, and that will only intensify as we have more products in line in the next year or two.

PR: Eventually you will be in this kind of discussion with a large number of groups, both big business partners but also companies from whom you need licensing?

VL: That is the part of my job that I have not really done too much of, just because we're still in the R&D phase. But it's the part of the job that I do fear a little bit, because that's where the stress will really come in, depending on how many markers you find in a product. When we talk about this constellation concept, I don't think any one of us right now knows how many stars would be in a constellation. It will vary from disease to disease, but if you're talking about six markers in a constellation that can predict Alzheimer's disease versus sixteen markers in a product that predicts heart attack—the amount of work that goes into figuring out what patents may impact the product—it's going to be much more work for the sixteen-marker product than the six. And so if, let's say, half of the markers are affected by third-party patents, you have eight patents out there, and they're owned by a collection of six parties, then you are potentially talking to institutions that may own them, companies like Genetech or Chiron that may own a couple of them or other smaller companies. And everybody will have a different expectation regarding the value of their patent, and some may want a 1 percent royalty, others may want a 5 percent royalty. That stuff is something that you cannot predict, and there's no way of really having a formula until we've done it a few times, and we will have to talk to all of the parties involved that have relevant patents that would affect the elements or components of our products. And it could range from 1 or 2 to 10, 20, 50 percent. It's something that we're very keenly aware of and we will need to face, but we also understand that there's no way that we can put out a product that is profitable if everybody demands a 5 percent royalty and you have ten such licensors of patents.

Stress

PR: We asked Joe Catanese if he finds his works stressful. Do you find your work stressful?

VL: Yes! I would be the first to admit that it is stressful. Because of the time-sensitive nature of a lot of these things, you can't afford to sit back. But this is not the only stressful job I've had. With what I do, wherever you work, it involves a fair amount of stress. It's just the nature of this business. If you're a really hyper personality, then it only makes things worse.

Boundaries and Collaborations:
Interview with Tom White, March 14, 2003

PR: How do you handle the issues of confidentiality, property and money in your collaborations with university researchers?

TW: I think your question derives from the Alzheimer's workshop that you observed. Since that was the first study that we undertook, the first big disease association study, it has its own intrinsic interests as a disease area, but it was also the vanguard model of how we would set up these collaborations. A number of the things that we're doing have never been done before, and we're using them both for the purpose of doing the study or making the test prototype but also to set up the procedures by which we will do all the other ones that will come after.

The diagnostic need in Alzheimer's turns on being able to diagnose the state of progression to frank Alzheimer's earlier than by just using cognitive measures. For people in their late fifties or middle sixties, with mild cognitive impairment—that is, more impairment than just forgetting where your car keys and glasses are—a physician would like to be able to have some more reliable method of knowing that this is really leading toward Alzheimer's. Today the disease is only formally diagnosed by autopsy of the brain. We decided that Alzheimer's was an interesting area because even though there are no good existing treatments, there are twenty different new chemical entities in the pipeline. So the issue isn't how you use the information for treatment today, but three to five years from now. If you could say these people are now on the path to Alzheimer's, you expect there will be better treatments available. Even more important, if those genes not only diagnose Alzheimer's but also indicate a risk for predisposition, then people would be treated earlier. Our model is AIDS patients with high viral load, who are now treated long before they have AIDS as a way of preventing progression to frank AIDS. We decided to study Alzheimer's for diagnosis, but maybe in five years, with better preventative treatments, there will be even more reasons to test people for risk.

PR: Have you found the functional SNPs?

TW: The functional SNPs we have identified might be causative, and therefore the genes might also be targets for new therapies as well or be useful to stratify the new drug treatments in clinical trials to indicate who's likely to respond. Maybe they'll be more effective in one subset than another.

We were interested in getting samples to find diagnostics markers; that was the initial rationale for exploring academic collaborations. Anibal Cravchik, an Argentinean, who is a Celera Genomics employee and an M.D., Ph.D., knew some of the Alzheimer's investigators. Three academic labs simultaneously published linkage articles in the December 2000 issue of *Science*. They all identified chromosomes 9, 10, and 12, and the linkage regions showed some overlap. Here was a good place to start looking at candidate genes. From December of 2000 through October of 2001, we looked hard at these studies. We encouraged Anibal to begin establishing collaborations. He is crucial, because he established a key initial collaboration, helped write the research proposal, and worked on the proposal that went to the institutional review board that approved the collaboration. We established two collaborations with university teams.

Then we identified a lab at UC San Diego, run by a clinician who was attending meetings and giving talks about the clinical aspects of Alzheimer's. We knew we would need his patient collection for replication. In the summer of 2002, we visited him. We talked about what we're doing, and he said, "Great, I'm a clinician, I know nothing about genetics but if you can do this, it would be very important to my patient population, and it will help me make the distinction. So if our collection can be of help to you, I'm happy to try to facilitate." It went through amazingly quickly. He is in the forefront of clinicians who are working with patient support groups to determine how best to use the clinical study of patients with severe diseases where the patients themselves may or may not be able to give informed consent, as well as where their family may or may not be able to decide in the interest of the family, who may also be at risk for a disease that has a genetic component. He told us during that meeting that they had been involved in drafting a law that had just been passed by the California legislature that said members of the family under certain compassionate conditions could give informed consent for people with severe neurological disorders. California's the only state that has passed such a law.

PR: Were the patient groups in support?

TW: All the patient groups were in favor. We had started doing the genotyping in June with the first hundred SNPs and the first eight hundred
patients and gradually working our way through them, because we
wanted to do a lot of patients with every group of SNPs because you
get statistical power from having a lot of cases and a lot of controls.
You could imagine doing it two ways: a lot of SNPs on a few patients,
but you wouldn't have any statistical power; or a small number of
SNPs on a lot of patients. Therefore, anything that you hit you know
you have done something significant, and hence you can roll over the
hits and try to replicate them while you keep slugging your way through
the other SNPs. That was the process we were using. Around the middle of October we began to get really significant hits, so it became
important then to get other samples for replication. We originally
thought we'd just send our hits to our chief university collaborator
and she would replicate them in families that she had, because the
family samples could not be transferred to us. The informed consent
for the family studies did not allow them to be sent to a third party,
whereas the cases and controls could. So we agreed that we would do
the discovery, she would do the replication. But then we began to realize that replication studies in families, for certain statistical reasons,
have lower statistical power. We wanted to check them in the U.K. and
in Europe and Japan, and we were going to need more big collections.
So the San Diego sample set turned out to be perfect timing, even
though their sample collection was in kind of an uncoordinated state.
They didn't report how much DNA they had and how well it was
quantitated, how clean the samples were, so we had to do a lot of work
to go down there, get their samples barcoded so that we could keep
track of them, barcode the identified patient records so we could eventually match the things, and then get the samples up here and make
sure they're right. And it progressed from there.

PR: How did you protect this? What kind of agreements did you have?
And where did the finances come in?

TW: Okay, initially when we did the analysis of what diseases we're going to
work on, we narrowed it down from 2,200 conditions to about 120
and then to 60 and then to 30, and they were analyzed in incredible
detail and we picked about 15 that are the top, unmet diagnostic
needs that meet certain scientific, medical, and business criteria. But

by and large we do not discuss publicly exactly what they are. Because if I say heart disease, it could still be predisposition, diagnosis, rate of progression, choice of therapy, avoidance of toxicity, disease initiation, severity, et cetera. For every one of the diseases, the most important question is not necessarily obvious and certainly not necessarily obvious to our competitors. In general, we say that we're working in six disease areas—central nervous system, cardiovascular, inflammatory diseases—but in some cases we have not yet said what is the actual disease that we work on, much less what's the specific diagnostic indication. But gradually, as we form these collaborations—for example, Alzheimer's is one of the CNS diseases we've been working on; in cardiovascular we are working with UCSF and Bristol-Myers Squibb; and in the inflammatory diseases we now say that we're working on rheumatoid arthritis. But in the other areas it's still not public exactly what we're working on. And even in rheumatoid arthritis, it doesn't say which of the eleven things we're working on. So parts of it are public, with people just assuming they know what we're working on, and most of the time they assume it's predisposition for disease, because that's what everybody talks about. And I can say that in most cases that's not what we're working on. But we don't feel any obligation to tell our competitors until we find something.

PR: That's all very clear. There are different publics—investors, competitors—although there are overlaps.

TW: So now the question is, when we're establishing collaboration, let's say with an academic group, we have to figure out, Well, how do we approach them? First we look for experts in the field, and then we approach them through scientific meetings or the disease area people or through personal contacts. And we decide, Do their research interests overlap enough with ours? Or do they have the patient population? Or a study that they're doing that would be adequate for the kind of study that we want to conduct? The first step is nonconfidential. At the next step we take a calculated risk, which is to say, either under confidentiality agreement or sometimes not, we say, "This is a specific indication we would like to study. Is this of interest to you?" And if it is, we'd say, "Would you be willing to consult with us in this area of disease as a first step in seeing whether we can perform an active research collaboration?" And some people would say, "Yes, I'd be interested in being a consultant." These are nonexclusive consultancies, because we don't mind if they consult with anybody else. But in the area of consulting, it's now specifically on that indication that is the one that we're working on. And in that area, if we tell them our confidential information, we expect they won't disclose it, but that's really the limit of it.

PR: Is that really enforceable? That's the part that intrigues me.

TW: Well, it's a confidentiality agreement or nondisclosure agreement. Victor Lee would say that those agreements are always enforceable. So our expectation is that the investigators treat the agreements seriously, but you have to be pragmatic about it too, because they have collaborators and people in their laboratory, and they have to discuss the research if they want to do the collaboration. But it's at least the start of asking, Is it likely that either they'll just provide advice to us, or that we can use that as a test of each other to see if we want to have a collaboration? The approach that they're taking is to guess at gene 1 or gene 2 out of a possible 2,500. Our approach is much more powerful, but it might be something that they still want to do; they would like the chance to discover the genes that are linked, and therefore it's of interest to them to collaborate with us.

The next step is an actual research agreement. It allows them to transfer the samples under the appropriate informed consent, with their own institutional review boards and the university approving it. Once that is done, it becomes a research agreement in which there is usually some funding we provide to the investigator's lab. For example, one collaborator already has some research funding of her own, but she said that she would like funds to buy the equipment to scale up the level of genotyping she could do (and pay for the reagents). We agreed that we would make that part of our contribution to her research. "You'll transfer the samples to us to study, and we'll exchange information on what we find."

PR: Are there examples where the nondisclosure agreement was violated by some professor who said something at a cocktail party, like, "How do you feel about this region?"

TW: Yes, I'm sure that there are; Victor can cite many cases where others have sued people for violating confidentiality. In my career, I know a number of investigators with whom we have had a confidentiality agreement, including the earliest ones in the PCR method, who went right out and used the Taq polymerase and published a letter in *Nature*, or claimed that they had done PCR on RNA, when it was very clear they heard that at our scientific retreats, and a whole slew of things where confidentiality has been broken by members of a scientific advisory board, much less collaborators. So it's always a calculated risk. You just have to make your own judgment, and you are careful as to the amount and degree of information that you provide, and you kind of assess how reliable the person is. And some people are more reliable than others, regardless of the confidentiality agreement, and you just are careful about that. If you have the sense the person is not reliable, you are just much more careful about disclosing the most sensitive information to him, but you still disclose confidential information. It's just less.

In this particular instance, until we began to find hits, the information that was most confidential to us was exactly how we were doing the studies, a specific way they are set up in our lab. Because again, there's Allen Roses at GlaxoSmithKline, with a $3 billion R&D budget; there is Sequenom; there is Athena—everybody else wants these genes. And there are academic competitors with whom we are not collaborating because we don't trust them. So we had to be careful as we went forward, especially when we began to find things. Because disclosing the name of the gene right away, somebody else could say, "Well gosh, I know the SNPs in that gene, I'll just test them myself and race into print with a correspondence in *Nature*, not a 'letter,'" like some of the things we've seen so many times before. We provided our Alzheimer's hits to our collaborators and said, "Remember, this is really confidential to us. These are the ones that are statistically significant in this population, so we don't want this to leak out, so please be very careful when you test these."

PR: So you have a combination of some assessment of their character, whether they're good people, but that could only carry you so far. Then you have the legal weapons if they do violate an agreement, and you find out about it in time. Finally, you both have some sense of mutual self-interest.

TW: They don't want to be scooped by the information leaking out any more than we want somebody else to announce the gene first or use it.

I think it's less the legal weapon than the notion that this collaboration is working. In fact, they have weekly conference calls. What you saw at the meeting was the getting-together part. The conference calls are among the most open I've ever participated in. On the call, the two labs talk about what they have done in the last week or two. We have a longer cycle, because the scale is 1,000, 10,000 times greater; so there's a delay, because we're making sure that the genotype calls are correct and that the statistics are right. So there's a loop in there to make sure that we don't make an error, because that would waste everybody's time. But the calls and transfer of the information—we're all in a race to discover genes for Alzheimer's, and they're our collaborators, so we disclose to them what we found so they can replicate it.

PR: So is your money involved at this stage?

TW: No, they get an annual consulting fee.

PR: Ten thousand dollars? A million dollars?

TW: Oh, no. Well, it's somewhere between eight and fifteen, or maybe it's twenty or something like that. And the total funding, I think, to one lab was on the order of $130–$150,000. We said, "What will support your research?" And she said, "I'd like to buy a $100,000 piece of equipment and another one that cost $20,000 and $30,000 of reagents." We said "Fine." That was it.

PR: And then where's the patenting in all of this? Who could patent what?

TW: So the agreement has a couple of key components. One is that she's continuing her research on making the best possible guess at genes that she thinks might be good to study on biological grounds. So we say, Well, she brings potential for sole inventions in her laboratory. If she makes a discovery, she and her colleagues are the owners, but we reserve the right to license it, either exclusively or nonexclusively, at our option—if they make a discovery, under the period of this grant, for which the funding is partially germane. Similarly, we may make a discovery using samples in a completely different collaboration that had nothing to do with the use of their samples, and that would be our discovery and it would be solely ours. But where we're using their samples in an open collaboration, chances are good that there would be a joint invention.

PR: But it wouldn't be solely yours if you did it with another group—it just wouldn't be hers?

TW: It wouldn't be hers, that's correct. You cannot agree on this in advance, because you can't agree on inventorship in a contract. Inventorship is a legal determination made by patent attorneys. So you say,
"Why can't I agree that we'll always be joint?" It's because it's against
the law. Attorneys make the decision as to what is the invention and
what did you contribute and what did I contribute or not. Most scientists confuse coauthorship with coinventorship, and they are totally
different. But in any case, in this kind of joint collaborative project, it's
quite conceivable that she might say, "Well, gosh, you know, within
this linkage region I'm going to study these three genes over here, but,
boy, if you could include these ten within your first hundred, I would
place my bets that those are really good ones to include." Or she could
say, "From my knowledge of the field, I'd start at the left side of the
linkage peak and I'd work left to right." Or maybe the replication data
could be considered a contribution to invention, even beyond the
discovery, or not. The agreement is written so that regardless of how
it comes out, we at Celera Diagnostics will be able to practice our invention, either solely Celera, joint with the university, or solely the
university, but we will be able to pick and choose whether we want
nonexclusivity or exclusivity. Because even in the joint invention, joint
invention means joint ownership, which means we can use it ourselves for whatever purpose we want, but we'd also have the option to
buy back the university's half-ownership so that we would then own a
joint invention exclusively. But we might choose not to do that. That
kind of agreement is typical in the philosophy of the approach, which
is to say, it could be sold either way, but we have the option to license
theirs exclusively or nonexclusively; or it could be joint, in which case
we're fine, but we have the option to pull it in exclusively. It's inherently nonexclusive because it's co-ownership. Most of the agreements
are written that way, and most of them are written that way strictly for
diagnostic rights, not for therapeutic rights, because we're only interested in diagnostic rights.

PR: How long are these agreements for?

TW: They are for the life of the patents, if any. The option period usually is
one year—one year from filing of the application. Or sometimes it's
on issuance, because you never know if they'll issue. But in principle,
the option period is usually about a year because you'd like to at least

see how it goes from the first rejection from the Patent Office, because they always reject the first applications. You want to see what their objections are. But that's how the research part of the agreement works. There may also be a materials transfer agreement that simply covers the transfer of the samples. Any materials that transfer back and forth almost always have to be covered by this agreement. That usually just pays for the shipping of the samples. It says they warrant that they have the right to send them to us and/or that they're covered by an informed consent that the university has sent to the IRB, because with these things you have to have belts and suspenders. You want to make absolutely sure they have the right to transfer those samples.

PR: Where would this kind of arrangement not be mutually beneficial? You are painting an awfully rosy picture.

TW: Well, I think the area that constitutes inventorship. And it's not absolutely clear in these things what constitutes inventorship, because you have to ask, Is just providing samples evidence of inventorship? Probably not.

MODELS ORIENT, TECHNOLOGIES PERFORM, SAMPLES SPEAK (OR VICE VERSA)

Celera Diagnostics has multiple disease groups. We hesitated about whether or not to include a narrative of one or more of these groups in our chronicle. Over the course of the first six months of 2003, the Alzheimer's group was the front-runner in terms of collecting samples and identifying SNPs. As we monitored this work, and that of other groups to a certain degree, the complexity of clinical detail and its interpretation was constantly before us as a narrative challenge. The challenge consisted in pondering how much technical detail would be understandable to a broad audience. We knew that intense debate within the company was taking place over how best to identify, evaluate, and, ultimately, present SNPs considered to be significant. In the light of this quandary, we decided to present the search for "models" and their relationship to the "sample" collections. Although this choice still meant including a certain level of technical detail, it had the advantage that the issues involved were general ones, engaging the overall strategy of the whole company, and could be discussed without revealing the specific goals or findings of individual disease groups.

Monogenic to Polygenic

For the first decade of genome mapping, the exemplars of the relationship of disease and genetics were the so-called monogenic conditions. These conditions were exemplary on a number of planes. They formed a model of what type of science was at stake as well as its clinical importance. This model was accepted among scientists (as they raced to find genes); it was the operative link between the scientific community and various publics (patient groups, funders, media, etc.). Thus, whether it was Huntington's chorea or cystic fibrosis in the United States or the muscular dystrophies in France, the model held that a small number of nucleotide changes in a gene—changes that mapping and sequencing would identify once the gene was found—were the locus of trouble. They were the physical place where small changes in the DNA (mutations, polymorphisms) produced significant changes in the coding for proteins—the products of gene action. These changes were responsible for the dramatic pathological phenotypic changes.

It was consistent with this approach that once these loci were identified, and the normal and pathological variations established, the path to therapy followed. The nefarious effects of miscoded proteins could be countered through one form or another of therapeutic response (genetic or otherwise), abortion, or, perhaps in some future time, genetic intervention before birth. Many books (or films) written during this decade of 1990–2000, while the genome mapping and sequencing were underway, were given the form of discovery or quest tales. These were stories of hope and commitment, or money put to good purposes, of dedicated (if competitive) scientists joining forces with patient groups, as well as either the government or the pharmaceutical industry or philanthropy, in search of a cause and a cure. Such stories moved from the unknown to the known; they had heroes; they had dramatic tension, because audiences and participants wanted to know who would win the race and appreciated stories in which suffering as well as fate could be supplemented with hope. In line with this genre of progress, inevitability, acceleration of events, and hope were numerous ethics conferences and debates largely devoted to imaginary futures and to laying out the universal principles that would protect the innocent. These principles varied from continent to continent, as did the stories told about them and the philosophic analyses produced to ground them.

By the beginning of the new millennium, however, a somewhat different understanding of how genetic action took place was beginning to take hold within the molecular biological community. One of the prime payoffs of the genomic mapping projects was the discovery that there were many fewer genes than expected, if by genes one meant what had been traditionally meant by genes (specific sequences of DNA, locatable on a chromosome and coding for specific proteins). Hence there were new challenges, among them how best to think about and to represent the new and more complicated—in the sense of having more steps to explain and more variations to chart—versions of what a gene, or gene action, actually was. One of the characteristics of these efforts was a certain conservatism of discourse; statements often did not explicitly underscore that significant shifts in the concept of gene action and the technologies used to experiment with it were taking place. Outside observers might be justified in wondering why all the money had been spent on mapping the genome if its most surprising results were so consistently downplayed. In any case, the complete sequence, the Holy Grail of genomic sequence data, the "Code of Codes," "life's program," turned out to be not the end of the quest for biological understanding but rather closer to the beginning.

Preceding this shift in the understanding of the gene was the prevailing narrative of discovery as a progression in understanding (and eventually therapy) that moved from the simple genetic conditions to the polygenic ones. The monogenic diseases or pathologies would be charted first; then the more complex task of the polygenic conditions could be attacked. Thus, for example, once the gene for Huntington's was identified, then the march toward the discovery of the genetic origins of more complex conditions—for example, the genes for Alzheimer's—would accelerate. In the early years of the millennium, within the scientific community, this narrative has been significantly recast. Outside of the academic community, the fact that something like 97 percent of the nucleic acids in "the genome" were not coding regions—their functions, if any, were basically unknown—was rarely mentioned after the first paragraph of the ever increasing number of books on the meaning of genomic research.

Exemplars and Models: From Diagnosis to Susceptibility

The emergence of new specialties is always of interest. Such a new area of expertise, bioinformatics, and its attendant specialists, grew in importance during the mapping decade as the quantity and meaning of the data being accumulated seemed to demand new technologies to handle it as well as new concepts to order it. New computer programs would scan the ever-increasing "mountains" of genomic data for patterns of genetic location and expression. This work was simply beyond the intuitive powers of individual scientists.

As we have seen, those working within the paradigm of linkage and linkage disequilibrium mapping operate within this rationality. The "hapmap" approach officially adopted by the U.S. government genomics community is a more sophisticated variant of this approach. There must be functional blocks of genes, and better bioinformatics would reveal them. Although these claims are reduced to a caricature here, they essentially represent the core logic of the dominant research strategies.

As we have seen, the initial strategy of Celera Diagnostics was not too far away from this logic. Celera Diagnostics had the unique advantage of having Celera Genomics behind it and drew the sensible conclusion that searching for functional SNPs in coding regions would be a productive way to proceed. Although that approach has been productive, it has not been sufficient, and Celera Diagnostics' strategy has evolved on a number of fronts.

It is striking, but not surprising given the primitive state of genomic and clinical knowledge, that a model for understanding the functional dimensions of polygenic conditions has only slowly been articulated—this, despite all the talk, for well over a decade, about its inevitability. It is worth emphasizing that there have been few mathematical models in molecular and cell biology (as opposed, for example, to population genetics). With the arrival of a large number of bioinformatics specialists, models with a strong mathematical component are now obligatory.[1]

Thus, at organizations like Celera Diagnostics, early in the new millennium, a corps of programmers, mappers, clinicians, high throughput technicians, and all the associated subspecialists that we have encountered was being assembled. But what was the best model on which to base diagnostic tests? This question—this problem—was very much under discussion at

Celera Diagnostics. In the spring of 2003, Tom White suggested that the article "Improving the Prediction of Complex Diseases by Testing for Multiple Disease-Susceptibility Genes," by Quanhe Yang, Muin Khoury, J. Friedman, and W. Dana Flanders, published in the *American Journal of Human Genetics* 72 (2003):636–49 and discussed below, was one of the more promising attempts to give coherence, and hence direction, to the approach to disease association studies.

ARTICLE ABSTRACT

Studies have argued that genetic testing will provide limited information for predicting the probability of common diseases, because of the incomplete penetrance of genotypes and the low magnitude of risk associated for the general population. Such studies, however, have usually examined the effect of only one gene at a time. We argue that disease prediction for common multifactorial diseases is greatly improved by considering multiple predisposing genetic and environmental factors concurrently, provided that the model correctly reflects the underlying disease etiology. We show how the likelihood ratios can be used to combine information from several genetic tests to compute the probability of developing a multifactorial disease. To show how concurrent use of multiple genetic tests improves the prediction of multifactorial disease, we compare likelihood ratios by logistic regression with simulated case-control data for a hypothetical disease influenced by multiple genetic and environmental risk factors. As a practical example, we also apply this approach to venous thrombosis, a multifactorial disease influenced by multiple genetic and nongenetic risk factors. Under reasonable conditions, the concurrent use of multiple genetic tests markedly improves prediction of disease. For example, the concurrent use of a panel of three genetic tests (factor V Leiden, prothrombin variant G20210A, and protein C deficiency) increases the positive predictive value of testing for venous thrombosis at least eightfold. Multiplex genetic testing has the potential to improve the clinical validity of predictive testing for common multifactorial diseases.[2]

The authors open by drawing a simple but trenchant distinction between genetic tests that determine whether a mutation is present or not (tests that have already been successfully devised) and tests that might predict "a healthy person's probability of developing a disease [. . .] of presumed multifactorial origin."[3] Critics from within the scientific and clinical communities have raised doubts about the feasibility of the latter goal because, they argue, merely identifying a gene (or SNP), even one involved in known disease processes, is insufficient; other factors, especially incomplete penetrance and overall low risk of a specific gene within a genotype in a population, complicate the task. Thus, while tests have been developed that accurately indicate the presence or absence of a marker for a pathological condition (for example, forms of breast cancer), their clinical worth is limited.

The authors not only accept these critical limitations but take them as their starting point: the remedy is in the evil. The problem is whether a model can be developed that overcomes the limitations of previous (essentially monogenic) approaches. Not surprisingly, the authors propose a conceptualization of "multiple predisposing alleles" as the place to look for an answer. Assuming that in many instances "if several factors (e.g., genetic loci) play a role in disease etiology, then, under many conditions, evaluating such factors concurrently (e.g., through use of a panel of genetic tests) substantially increases the predictive value for the disease."[4] The reader should notice the cautious beginning signaled by the use of the word "if," as well as the conclusion which can be read to suggest that the path forward involves steady progress rather than an entirely different understanding of disease conditions and ways of modeling them.

The logic of the model the authors propose is straightforward. Where the logic meets the science, as it were, is in the assumptions made about clinical data. In the year 2003, such data are scarce at best, and not validated experimentally or clinically. As far as Celera Diagnostics (and other research teams) are concerned, it is strategically compelling to radically change this state of affairs in very short order.

The model rests on what is referred to as a "likelihood ratio." The likelihood ratio "reflects the probability that a patient with the disease has an observed test result, compared with the probability that a patient without the disease has the same result." Thus the challenge consists in building a model that will increase the significant difference between test results to a point where they justify clinical screening and diagnostic testing. The older form of genetic testing asked simply, Is the marker present or not? Tests based on the new model would ask, How likely is it that this particular person will develop the disease? "High-risk alleles at any single locus often occur in persons in which the disease will never develop, and low-risk alleles often occur in patients in whom the disease develops. According to the multifactorial model, the disease will develop only in people whose combined burden of genetic and environmental risk factors exceeds a certain threshold." A test for predisposition must be capable of taking all these factors into account: age of onset, environment, hereditary factors, and others.

The idea of developing tests that combine information from multiple risk factors to establish probability has been explored, for example, with breast cancer, "factors such as age at menarche, age at first live birth, number of previous breast biopsies, and number of first-degree relatives with breast cancer."[5] The current model is more sophisticated in its ability to bring more genetic variables into the equations. As data accumulate on population frequency rates of alleles, the model will gain in efficiency and power. As more is known about disease incidence in specific populations, "common multifactorial diseases might be more reliably predictive than conditions that are neither common nor multifactorial. This view is supported by the fact that a positive panel of tests for common alleles and relatively weak risk factors, when taken as a whole, may be as informative as testing for a single, strong risk factor."[6]

The ability of genetic tests to predict multifactorial diseases is not inherently low but depends on how many factors are considered and the characteristics of each factor with respect to population frequency, associated risks, and interactions. As knowledge of these factors and their associated parameters improves, so will the ability to predict the probability of developing a particular disease. The major limiting factor might be the background risk in the population. The simultaneous presence of genotypes that confer a lower risk adds complexity to the scenario but can easily be included in the calculation. Such considerations are valid to the extent that the model implicit in the test panel correctly reflects the underlying etiology of the disease.

The authors conclude by reminding the reader that the validity of the method ultimately depends on the data that is used. The caveat applies not only to each of the risk factors taken singularly but also to each of their interactions. "If such data are lacking, estimates of summary risks would be incomplete and possibly misleading. Unfortunately, however, such data are lacking for most conditions."[7] The article's authors call on the clinical and epidemiological communities to produce this quality data.

Bringing it All Together: The Cardiovascular Group

We shared with Tom White, from time to time during the course of the spring, our hesitations about whether or not to choose one disease group to include in the chronicle. One day he suggested, with a certain enthusiasm that apparently sprang from recent discussions over strategy, that he found Jim Devlin, head of the cardiovascular group, to be an especially thoughtful and articulate individual who was balancing adroitly the uncertainties and complexities of the situation while nonetheless continuing to move ahead decisively. We approached Devlin about doing a series of interviews; he checked with White, and then agreed.

Interview with James Devlin, Ph.D., June 19, 2003

JD: I received a Ph.D. in biology at the University of California at Irvine. I was in an immunology lab where we were purifying proteins from white blood cells to see what effect they had on the immune system. It was very frustrating, because we would take cells and stimulate them with an impure stimulant, then take the proteins and try to purify them, then take a partially pure protein and put it on a partially purified cell preparation and get an extremely confusing result. [*Laughs*] I decided that the way out of this was to learn molecular biology, so we would know how to clone the proteins and at least have pure proteins for experiments. So I did. While looking for a postdoc position, I ended up through a chain of circumstances at Biogen in the Boston area, Cambridge. This was the early '80s. We were working on the HLA (histocompatibility complex antigens). It was fun, but I missed the Bay Area, where I am from, so I took a job at Cetus in 1985. When Cetus split up, I went to Chiron. Then I was feeling adventurous, so I left to try to start a company with some academic folks focusing on immunology and display technology and raised some seed funding from Johnson & Johnson Development Corp. but could never quite get enough people interested to get it off the ground. I went to Berlex BioSciences in Alameda and worked on small molecule drug development projects. I led some technology groups—bioinformatics, protein biochemistry—that supported the disease areas in Berlex, such as cardiovascular. I stayed there for seven years. When I heard that John and Tom were thinking about heading off on this adventure, I entered into discussions and joined John's group in March, right after Celera Diagnostics got going.

I spent the first year here with Joe Catanese, focusing on getting the genotyping infrastructure up and running. I also spent a good amount of my time working to obtain access to clinical samples for association studies. From the beginning, my concern was about how hard it was going to be to get the samples to do the studies we wanted to do. The genotyping infrastructure is working pretty well now, and we're about to start these studies.

PR: Can you tell us how the conception of the platform has changed over this time?

JD: Well, when I first came, the idea was that the disease groups would decide what SNP assays they wanted to build in order to look for disease associations. They would build as well as validate those assays for their favorite candidate genes. They would start getting samples, and they'd do some of the small association studies themselves: they'd genotype them and see if there were associations—it would be kind of a collection of little cottage industries, each doing their own thing with racks of test tubes. That's the way big pharma typically did drug discovery. If they wanted to do something, they'd collect a group of chemists and they'd start making compounds and testing them to see if they worked. But the industry had already started to change and was assembling very large libraries of small molecules and screening them.

PR: Was this rational drug design?

JD: No. This was irrational drug design. I would not say mindless, because I was involved in writing a computer program to improve it, but it wasn't very rational, just big. At Berlex I worked with the chemistry groups, and one of the first things I did was lead a project to set up the first high-capacity screening program. I negotiated with all sorts of folks to buy large libraries of compounds. We needed to assemble a huge library, assemble the robotics to deal with the library, develop a bar code LIMS-driven tracking of the library, so that we could take our assays and run many, many samples through them in an anonymous fashion without worrying too much about any single sample. The logic was, if we take enough through, we'll get some hits. Joe Catanese came from a similar background. In the initial vision, I think, Joe was going to do some of the overflow work from the disease areas. I thought in order to have many, many validated assays, we should develop an automated design system—design many, many assays and rather than have each scientist handcraft a particular assay in a very loving fashion, we should just design them all in bulk and run them through the system in bulk, which would be efficient and fast. The assays that failed, well, we could either redesign and rerun through the automated system or, if worse comes to worst, have people lovingly handcraft them. At first, there was a great deal of resistance to this approach because people were concerned that "these are really important assays for our disease area. I don't want somebody else doing the validation experiment who doesn't really care about these assays and isn't invested in them." Joe and I told people, "Well, that's okay. All the trays will be unlabeled, so no one will know whose assays are in the rack. So they could be their own assays, and by the way, all anybody is doing once the assays are in the rack is moving a rack of tubes from one robot to another. What can you do wrong as long as there is a bar code label?" We worked at it and people came to see that it was really the only way to do what we needed to do in order to get the large number of assays validated that we needed validated. Initially, we were using small robots, and we had the disease area folks actually manning those robots, but once Joe got his Packard plate track, we were able to switch from those small robots to the plate track and use Joe's dedicated team. It's absolutely the most efficient way to go. So that's what changed. We've spent an enormous amount of money on the high throughput infrastructure, which we planned to use for the association studies, but we've used it very effectively to validate assays.

PR: Did the amount of money that Celera Diagnostics has available make this possible for you?

JD: Certainly it made it possible. But that wasn't the challenge. Elsewhere, that was the challenge: getting the money to acquire the robotics and infrastructure. Here, management had already done that part. What we really needed to do was help the disease area folks feel comfortable with saying, "That may have been the way you did molecular biology before, but if you want to validate 20,000 or 30,000 SNP assays, you need a factory—an army." I would say the inspiration came not from Celera Genomics but from the public genome project—that's where Joe came from—and from pharma industry and the high-capacity screening, because—that's where I came from. Joe and I pushed to take advantage of Joe's system.

PR: How long did it take to get the assay system going then?

JD: Well, you say that as if it's going [*laughs*]. It started when we started, and we're still making it better all the time. There were several things that needed to come together. One was the assay design system. Another was the database of SNPs that we wanted to design. That was a project in itself. And then there was designing the robotics infrastructure. There was also the choice of which enzyme we would use, and there were two competing candidates. We spent the first eight or nine months debating this issue, and it wasn't until January or February, I guess, that we finally decided on which enzyme we would use for PCR. So it was really after February 2002, that we started the official assay validation process using the robots. Then shortly after the plate track arrived, we switched to the plate track system. And then we had another significant improvement in the plate track system probably around last Christmas. So it's been a continuous process.

PR: And while this is going on, the collection of samples is happening as well?

JD: Yes.

PR: So you had time while that main task was undertaken to really work all this out?

JD: Exactly.

PR: Shall we skip over the enzyme wars?

JD: It actually was quite interesting, because there were two possible enzymes; two different groups were assigned to optimize the two different enzymes. Over the nine months we spent at it, it was leapfrogging: one leapfrogging over the other. We probably ended up with a better system because the two groups became competitive.

Samples, Predispositions, Markets

JD: We have a number of sample sets. We have a big sample set from UCSF from cardiovascular patients—UCSF, with a good deal of foresight, saw that it would be good to include broad genetic research consent when it was collecting samples. We have access to a sample set from Bristol-Myers Squibb from the CARE Study, a secondary prevention study; all of those folks had a myocardial infarction as an entry criterion for that study. So we have that large sample set.

PR: Large sample meaning?

JD: Three thousand samples, 5,000, depending on how you count, and 4,000 to 6,000 samples from UCSF. And then we're collecting samples for a thrombosis study now, and we have just signed an agreement for another heart disease sample set from the Cleveland Clinic—1,500 samples—and we're in various stages of negotiation with, I would say, eight to ten other sample collections.

PR: Huge.

JD: Well, you don't need samples just for the sake of having them; you need samples that allow you to answer a specific question. We're trying to develop diagnostics, and here we'll speak primarily of genetic diagnostics: we're looking for genetic risk for a particular disease. Since your genes don't change over your lifetime, looking at someone's genes doesn't really let you see whether or not you've just gotten the disease or how the disease is progressing. So looking at the genes can really just help you say, "You're predisposed to disease," or "You're predisposed to have a bad reaction to a particular drug," or "You're predisposed to respond very well to a particular drug for a particular disease."

Hence, in genetics we're talking about predisposition. What questions do we want to ask about predisposition? Well, since we need to do something we can sell to fund our continuing research and be viable here at Celera Diagnostics, we need to find a large, unmet need. We look for a disease risk and then ask, How large is the market for that particular disease and disease risk? And then the next question is, If you were to successfully say to this group of people, you are at enhanced risk of getting this particular disease, is there something useful they can do about it? If there's something useful that they can do about it, then they're going to want to buy and use your test, because it can help them decide, depending on the result, whether to act or not. If there's nothing they can do about it, then what good is the test? And no one wants to buy it.

So in the case of cardiovascular disease, for example—let's just take myocardial infarction or heart attack—it's an attractive diagnostic target because it's a very prevalent disease: heart disease is the leading cause of death in the United States and most developed countries. So there are a lot of people at risk of the disease, and if you know that you're at significantly higher risk than the general population, then there are quite a few things you can do about it. And more importantly, the things that you can do about it are relatively safe. If, for example, you had a disease where, if you knew you were at high risk, you could take this particular medicine or undergo some treatment that is very risky or has very many serious side effects, then you'd want to be really, really certain you're going to get that disease before you undergo the treatment. If, on the other hand, the treatment is something as innocuous as not going to McDonald's, or going jogging every day, or losing thirty pounds, or going on some cholesterol-lowering agent, then the barrier you have to surpass, as far as the precision of the diagnostic goes, is much lower.

So it seemed like quite an attractive market. We tried to find large populations of people who have had a heart attack—ideally, at a young age, which would indicate an enhanced risk for coronary heart disease— then compare them to a control population. Now, that's an interesting story: what's a control population? This question has spurred many vigorous debates. If you want to publish something in the literature, a good epidemiologist will tell you that the control population should be absolutely identical to the case group, but for the fact that the case group has the disease and the control group doesn't; that way, the only difference between those two groups is the disease, and your interpretation of the results is not confounded by any other differences. Well, if your objective is to publish something in a good epidemiology journal and not have official epidemiologists laugh at you, it's probably a good approach to take, but after all, our objective here is to develop successful diagnostics. To me, the problem with that approach is that a lot of people have cardiovascular disease; if I take a bunch of thirty-five-year-olds who have had a heart attack, which is an unusual event for people that age that suggests a severe risk for cardiovascular disease, and if I take as a group of controls some thirty-five-year-olds who haven't had a heart attack, well, what percent of those guys are going to have a heart attack over the next ten or twenty years? Probably a good number of them, because cardiovascular disease is a very prevalent disease. You can only have so many samples, because samples are expensive, and my bias is, let's stack the odds in favor of discovering something. Once we discover something that might be right, we can always go and check it, but if we don't find it to begin with, we have nothing to check.

So I said, let's compare a group of young people with cardiovascular disease to a group of people who are just the same but who we know will never have a heart attack in their life. And it seems like a good idea—but how do you know when someone will never have a heart attack in their life? Well, you take ninety-year-olds who go to dance class every day. Because you know they've never had a heart attack. So you take an old, healthy control group. It seems to me it gives you the best odds of finding something. Well, of course, this horrifies the epidemiologists and statistical geneticists, who say, "But you might just discover a SNP associated with longevity!" And I say, "That's okay! We'll publish that!" Okay, I mean, they're absolutely right. Some of the things that we're going to find are probably going to be associated with longevity, and that will be fun, but some of them will probably also be associated with cardiovascular disease.

So this is the first study that we're going to do—an exploratory study. We're going to do replication studies. We can't do everything to please everybody as we go forward. The important thing to me, though, is not to miss an opportunity. Usually, when you're trying to get an experiment to work, the hardest part is to get some glimmer of a signal. Once you have a glimmer of a signal, you can modify conditions to optimize it and improve it and watch the needle going up as you change things. But if the needle is never moving, you don't have any way of telling whether you're making things worse or better. So, to me, taking the young, sick people and comparing them to the old, healthy group is the best way to get the initial signal. As a matter of practicality, in order to avoid spending the rest of our lives arguing the point, I said, "Listen, we'll just do it both ways. We'll get the case group, and we'll get a control group that you like, and we'll get a control group that I like. We'll do it both ways and cover all our bases." And I should say that Tom and John liked the idea of the older control group, and they were fully supportive of doing it that way.

PR: Okay. You set out to get those samples, but you also already had some indication of what genetic markers or genes were involved, because a lot of research has been done on this already. How do you decide what markers you're looking for, and how does being at Celera change that process?

JD: There are three kinds of markers: There are markers that are already published as likely to be involved in cardiovascular disease. Of course we look at those. Then there are other SNPs that are in the public databases that haven't been thoroughly investigated. We of course look at those. Then there are SNPs that were discovered in the resequencing project. So of course we look at those. They are very exciting, because no one else in the world, at this point, can look at the data that came out of that; this is being in the right place at the right time. Even if somebody else has set up a genotyping infrastructure that is as effective as ours, they can't look at those SNPs because they don't know they exist.

There is a great deal of concern in the disease area groups that not everyone's favorite candidate gene has a functional SNP in it. Let's say a disease research group has a thousand candidate genes, which have been lovingly culled from the literature, and they know that these are the thousand most important ones because they've been published in the literature. Maybe only half of them have a functional SNP in them. Well, the thought of not interrogating the other half of the candidate gene list is very dismaying to people. People would like to look at LD [linkage disequilibrium] SNPs, as we'll call them for short, in those genes. But the advantage of a functional SNP is it's likely to be the causative SNP, whereas with the LD SNP you're hopefully close to something that's causative. Even if you do get a hit, you have your work cut out for you figuring out what the real causative gene and functional SNP are. But then how many LD SNPs do you need to add to the functional SNPs? Well, the more the better, so let's say five at least. Well, we're running at capacity here. We're putting as many functional SNPs as we can through our factory. Now, if you start saying, "Well, for half my gene list, I want to add five LD SNPs"—and, you know, a lot of those won't be valid, so it's going to be more than five per gene—those are going to knock functional SNPs out of our system. And, to me, it just didn't make any sense at all, because I would rather go with functional SNPs in genes that probably cause heart disease that no one has quite thought about and published, than go with LD SNPs in other genes that maybe people think are important. Sure, we're going to miss something, but if you get a hit with a functional SNP, it's likely to be on target. If you get a hit with an LD SNP, it's just the beginning of the very long-drawn-out process of pursuing it.

PR: I've done some work with deCode; they have all the advantages of the linkage approach. They have a lot of linkage regions but not too many jackpots yet.

JD: John has done an admirable job of holding the line to give absolutely top priority to functional SNPs, against a barrage of criticism from people who were horrified at the thought of not interrogating their favorite genes with LD SNPs.

PR: Is there a particular disease group that has suffered the most for this?

JD: Well, I think it doesn't even go by disease group. One of the scientists in my group is appalled and horrified by this approach. I know that the inflammation group and other groups are very much in favor of LD SNPs. I would say, other than myself, all the disease group heads were strongly in favor of the LD approach. [*Laughs*]

PR: So, at least within your group, you won the battle. You were going for functional SNPs. Having decided you were going to do functional SNPs, was it hard to decide which ones you were going to look at?

JD: No, we looked at every one we could. I think we have on the order of over 20,000 validated assays now.

PR: Is that enough?

JD: Is it enough to find something? Probably. Is it as many as we want? Probably not. Twice that many will probably be something like enough.

PR: Forty thousand functional SNPs? That's the number of genes in the genome.

JD: Yes, but just because you have two in one gene doesn't mean that one is good enough. Because they may be so far from each other that having a hit in one doesn't tell you about the other one.

PR: Are you looking for patterns? Or constellations?

JD: We're looking for prediction of risk. And risk could be predicted by a single SNP, if that SNP has a strong effect, or a combination or interaction of multiple SNPs. For example, if you have a village and a village has a drought, you know, it's kind of hard on the village. If there is a horrible windstorm, that's kind of hard on the village. And if someone's house catches fire, that's kind of hard on the village. But if all three of those things happened at the same time, you have ashes. So I believe you can have SNPs that by themselves have a marginal effect but in combination can have a very severe effect, and the only way to discover that is empirically. You can make mathematical models to take these individual SNPs and consider their individual effects on risks and mathematically model what the combined risk will be, but you have to make assumptions about how you combine those risks. Let's say you have two SNPs, each of which doubles your risk, and if you have both of those together, you might have a tenfold increased risk, or your risk may still only be twofold.

Models and Samples

PR: So, models are good things to have, but ultimately significant results depend on the quality of the clinical and scientific data, right?

JD: That's fine and that's true. Statistical geneticists would say that we assumed a kind of a multiplicative model with the multiple SNPs. We assumed that a twofold risk and a twofold risk, if you multiply them together, give you a fourfold risk. Let's say you have three dials, each indicating risk: if one dial is wind, and one dial is drought, and the other dial is fire, you multiply the readings on those three dials together and you come up with a very high combined risk. But let's say you have two dials, and someone tells you, "Well, each one of these dials predicts risk of fire and so you probably want to multiply the readings together." But if you pull the curtain aside and peer behind the dials, you might find that each of them is connected to a thermometer. One gives you a reading in Celsius and the other Fahrenheit. Well, they're both really measuring the same thing, so they're not two independent risk factors that you should multiply together. Any mathematical model assumes that the things that you are measuring are independent from each other. Yang said, if everything you're measuring is on the same pathway, then you are not looking at independent measurements and it doesn't make sense to multiply them together. But even if they're on different pathways, you still don't really know how to combine them. You can make beautiful mathematical models and you can say, "If this is the right way to sum the risks, this is what the result is." That looks nice, and the math is all right, but is that really the right way to sum the risk? The only way you can determine that is empirically.

The approach I like is to take all of the results we have and not only look at associated risk for each individual SNP but also look at them in combination. This is easy to do computationally even on a PC: Let's say you have SNP *a* and SNP *b*, and the relative risk of SNP *a* is two, as is the relative risk of SNP *b*. What if you have SNP *a* and *b* together, is the risk also two? Is it four? Is it ten? It could be any of those; you don't know. You could do math and try to guess, or you could make up some formula, but what counts is what's going on in the real world. Well, we have the data. So, instead of doing a standard statistical analysis with a two-by-two contingency table, you just say, "These are the people who are diseased, those are the people who are healthy, and these are the people that have SNP *a*, and these are the people that don't have *a*." From doing this you can get the odds ratio.

Let's say the odds ratio is two, and if you did a similar sort of thing for SNP *b* and got an odds ratio of two also, well, that's what our statistical genetics group does all the time. The other thing you can do, which drives the statistical genetics team crazy, is you can say, "Well, rather than calling SNP *a* individually and looking at SNP *b* individually, let's bundle this thing up and call *ab* one unit and treat it like a single SNP." So you make all possible combinations of two SNPs, for two reasons: Number one, any SNP taken individually is relatively unusual. If you multiply two unusual SNPs together, you get an even lower likelihood of co-occurrence, and if you add a third, well, you reduce the frequency even more. It's pretty rare to find somebody with that combination of three SNPs, but two SNPs you can still find quite frequently. So now you can compare the people who have this artificial double SNP to the people who don't. Now if you do the same thing, you get a real live odds ratio. So rather than saying, "Well, I've made up a mathematical equation that suggests something here is true," now we can say, "I know this combination of SNPs exists, I know that combination of SNPs exists, and I know that gives you something significant because here's the data: I did the experiment." But you're doing this on an empirical basis. You are not making up anything about the relationship or interaction of the SNPs; it's model-free, it's based on data.

So what's not to like about this approach? Well, the major concern of the statistical group is multiple testing. The more single SNPs you have, the more double SNPs you have. The statistical genetics folks will say, "Well, listen, if you make enough combinations and test enough things, you're going to find significant odds ratios and significant values just by random chance. You really shouldn't do that, because you're killing yourself with multiple testings." Well, the way I think about it is, if one of these double SNPs really is significant, then the fact that you've done lots of multiple testing doesn't somehow magically make it vanish in the real world. Out in the real world, it's still doing something real. Sure, if we do this, we're going to get a lot of false positives, but we'll just rank them. Take the top of the list into a replication study and anything that is real will reappear. Replication is the way; you can be frozen into immobility by a fear of multiple testing, by thinking you don't want to do many tests because then you will have to mathematically discount your results. The way I look at it is, Hey! We have the data. Let's do this analysis, rank the results, and take the best into the replication study.

PR: Can I ask you a naïve question? If I understand this right, eventually you end up with something like types?

JD: Yes. Now when you do a test, instead of saying, "Do you have a mutation?" you'd say, "Do you have the combination of these two SNPs?" Which is a very bad thing to have. And those two SNPs, by the way, may be on completely different chromosomes.

PR: But that combination is taken to be a unit?

JD: Yes, that is a unit. So the other criticism of this is, well, gee, when you start multiplying these together, you're going to have a very small percentage of the people who are going to have this, because you're multiplying two small things together. Not necessarily. If you take two risk alleles at a frequency of 20 percent and you multiply those together, now you have a double SNP with a frequency of 0.04. That's a pretty high-frequency risk allele, and that can have a very significant odds ratio—much more significant than you would have seen with either of these. We occasionally have a consultant—I don't know why we have this person—who comes through and says, "Look, you guys are out of your minds. You're never, ever going to find anything useful, because any SNP that has a significant effect, a really significant effect on a disease you care about, has already been found in a linkage study, okay?

Because, if it has a strong effect, the people doing gene mapping would have found it." Well, that's true if it does by itself, but if the SNP you're looking for has two components, and is scattered across two different chromosomes, no, you're not going to find it that way.

Process: Interview with James Devlin, Ph.D., July 3, 2003

PR: One of the main things we would like to concentrate on today is process. We showed Tom some of what we've written recently about other dimensions of Celera and he said, "It's good, but you're missing the fact that we keep changing our mind and we keep rectifying. Even if the goal is to build this stable and coherent diagnostic system, which will be disseminated widely all over the U.S., how we get there is a process of mass trial and error." Can you tell us about the process your group has gone through?

JD: Okay. Let's think about it: Has it gone as expected for us? And if not, what was different? Well, when I first started talking to John about joining him at Celera Diagnostics, I remember we went out to breakfast and I told him that I thought the key issue seemed to be the ability to assemble a well-characterized set of clinical samples. What surprised me was two things: number one, how much money it ended up costing to get access to these samples, and number two, that the company was willing to pay for it.

PR: [*laughs*]

JD: So from that point of view, the financial resources, I have to say, are great. On the other hand, it was still as problematic as I expected to get the samples because, in part, the company was enmeshed in a number of legal discussions at the same time. The alliance with Abbott took a great deal of our legal group's time and attention. Hence, the rate at which we could acquire access to samples was limited by the amount of legal time available here. I come from a therapeutics background, and I was surprised to discover the number of samples that were already out there adequately consented. So the samples were there, the money was there, and I think we made good progress at getting them. It took us a while to get our genotyping infrastructure up and running, so, yes, maybe we could have gotten started a little bit earlier, but not too much earlier.

PR: Two months?

JD: Yes, maybe, something like that. Because in some ways the sample sets became available at about the same time that our genotyping infrastructure was ready to deal with them. Things were reasonably well matched. I guess one thing that surprised me a little was the informatic aspect of it. I came from a medium-sized pharma company which had a reasonable information technology [IT] infrastructure, and when we set up a high-capacity screening system there, we were plugging into an existing database for handling our compound collection; whereas here we were building everything from scratch. So it has required a reasonable amount of effort to deal with a database and information management system that's changing on a daily basis. You're not in a stable environment, and it makes it challenging. But the one thing that I think has allowed us to be as successful as we have been is that the disease groups and IT itself all report into John Sninsky in discovery research. He's done a really great job in making sure that everybody is focused on getting the same thing accomplished. John's able to keep so many very different groups all focused in the same direction—it's amazing. His ability to do that is certainly what's allowed us to start from a blank slate and make such rapid progress in getting it up and running. We are trying to build a generic machine where we can ask any question we want, pour something into the funnel at the top, and out comes an answer.

As for the samples: Are the patients really all diseased? And if they are really all diseased, do they really have the same disease, or do they have several different diseases that look phenotypically similar? That's a problem. And your controls: Are they really all healthy? And if they are healthy, what does healthy mean? Does healthy mean that they're doing fine today? If the definition of healthy is that someone never has a heart attack, then you can't tell whether or not anybody is healthy until they die of something else. If they die of something else when they're young, maybe they would have had a heart attack if they had lived longer. So that's an issue and that's why I favor older controls, but that's a very contentious issue. Even for genotyping, there are major problems with the sample sets. You could say, "Well, let's find the perfect sample set and the perfect set of assays and do the perfect experiment some time in the dim distant future," or you can say, "This is the best I have today for samples; this is the best I have for assays. Let's do the experiment and see what we find, and if we find something exciting and interesting, we'll run with it." Will we miss something? Sure. But our objective is to find things, not to avoid missing things. The former is achievable. The latter is not.

PR: And you can always go back and enrich your findings and the diagnostic tests?

JD: When you said at the very beginning of this that the goal is to develop a standard assay set—well, no, that's not really the goal. The goal is to develop something valuable quickly, and then keep looking and keep adding to that initial set of assays or constellation of SNPs in order to improve its value as time goes on.

PR: Okay, but there is going to be a point in the not too distant future where you collectively have to submit some test to the FDA and to make agreements with the reference labs—that's the test for you.

JD: Certain tests, such as Analyte Specific Reagents, must be manufactured under GMP [good manufacturing practice] but don't come with clinical claims and don't require FDA approval. They only require a lab to validate a homebrew test and for us to be able to convince the reference labs that what we want them to buy today is better than what they bought yesterday. We need to be able to make the case to them: "Here we are today. I know this isn't what we were selling you yesterday, but it's better than what we were selling you yesterday, and you'll be doing better than your competitors if you start selling this."

PR: Don't payers require some stability?

JD: Yes, but the thing that gets companies to buy your test rather than your competitor's test is the claim that yours is better. I believe as it improves, you're always going to want to be going back to the reference labs and saying—I'm not in the business group but certainly from a scientific point of view—"For the five-year time frame, maybe ten, the information you can get from this assay panel will be improving year by year as the collection of markers in the panel improves and as our understanding of how to interpret the results from those markers improves."

PR: But you're not going to have to retool your lab. And you're not going to have to buy new software every three weeks.

JD: Okay. Now, what do we actually take to the FDA? It's expensive to take something to the FDA for clearance as an in vitro diagnostic kit; we'll take it to the reference labs first and see what kind of customer uptake the test gets. That will guide us in thinking about what to take to the FDA in order to get a kit approved.

PR: If you could, go back, rewind the tape of history and tell us when you felt you had your first constellation of significant SNPs. When did that become a reality?

JD: Well, the question is, When is that becoming reality? I don't think we're there yet. I'm an optimist. I think this has a good chance of working. I can't point to any constellation and say this is going to be the key to our success in the future. I think we've put the pieces in place to find out if the really useful constellation is out there. We've put the tools in place to give us a really good shot at finding it. We're just in the process of winding through discovery studies now. We're starting to replicate some studies. It's clear what we need to do. It's clear that we needed to get the samples. We needed a high-throughput genotyping infrastructure, and that is now in place. Then we start producing results.

PR: What is the number we are talking about? thirty?

JD: More.

PR: One hundred fifty?

JD: Five percent of the number you test end up with a decent odds ratio and p-value. The more you test, the more you get. You've collected a sample set, and let's say you have four hundred cases and four hundred controls. We have many, many hits, and we don't know which is the real one. Which is the SNP that you want to pursue? That's the challenge. How do we deal with the multiple-testing problem when we get all these false positives? One solution is to increase the sample size and look at replication studies, but can you really increase those numbers tenfold? If you find something with an odds ratio that's interesting, the fact that you've done a lot of multiple testing doesn't mean that this is not real. It's still real; it's just that you have all these false positives.

How do you deal with those false positives? Let's say we've done a discovery study with four hundred cases and four hundred controls, and now we're going to go on and do a replication study and we also have four hundred and four hundred. Let's just keep it simple and think about single SNPs rather than double SNPs. Let's say we had 10,000 markers we looked at. How many do you take forward? Do you take 10 percent of them? Your top 10 percent forward to the next level? You still have a significant multiple-marker testing question here. Where do we draw the line, and what do we take forward into the replication study? And my response to that always is, "Don't waste time worrying about it." The line is not going to be drawn statistically. It's going to be drawn by John Sninsky saying, "You can have this many days of plate track time or this many," and that's how you decide. For example, in the small replication study my group is doing, we've been given the time and space to do 96 markers, so we're going to take our favorite 96 markers forward into the replication study. And what are those markers? Those are going to be a mixture of things with very attractive-looking p-value and odds ratios and also some representing our favorite genes from a biological point of view. So the statisticians are comfortable with us, or maybe they just don't object because we don't tell them about our biological-favorites point of view. I mean, it's not just statisticians. Many people feel that our knowledge is so limited that maybe, in a small number of cases, our biological knowledge may help us prioritize markers, but not in most cases, not statistically. Occasionally we may be able to use the biological knowledge, but I don't think anybody really thinks it's a general method that will work in most cases. I think it's a great idea to incorporate the biology, but I also think that it's only going to work occasionally, when we happen to know something.

One solution to this is using our knowledge of biology to try to work through it. Maybe expression analysis in some cases will help with this problem. To many people, it's not a very satisfying way forward, because the amount of things we don't know is much larger than what we do know. So you can say, "You're going to miss a lot," and, again, my response is, "True," but we're graded on what we've achieved, not on what we've missed. I don't want to gloss over the problem: this multiple-testing issue is a major challenge facing us at this point.

PR: Your description makes this sound like physics to me, in the sense that the experiments you can do are a function of your place in the line and the amount of time on the machine. And biology has not been done that way before, so this is what big biology looks like.

JD: You need to do everything in Joe's shop in terms of screening and discovery. I would say the first wave of discovery data has just crashed down upon us and we're struggling to the surface, trying to figure out how we're going to deal with this avalanche of data, how we're going to analyze it and start doing replication studies. We're just getting data from our first replication study; we're in the process of comparing the replication and discovery results. We have some hits that seem to be possibly replicated. At first, it's just listing things that show up in the discovery and replication, and then you see things that repeat, and then you say, "Well, okay, that looks interesting but I don't believe any of this. Please go back and check the calculations." And then you go back and some problems come up because, again, we're doing this for the first time. But something still might look interesting. "Okay, the calculations are correct, but I don't believe the numbers that were fed into the calculations. Let's go back and look at the actual growth curves, the PCR reactions." So you go back, and you find some things that fall out when you double-check that step as well. We're in that process now of checking and double-checking things that seem like they've possibly replicated.

PR: And in September?

JD: I hope we'll be further along that path. We need to have several decision-making groups. We need to look at the replication and discovery results ourselves and think, "Is it worth pursuing this and putting the resources into starting to build a multiplex, or acquiring more expensive samples for validation studies, or funding external validation studies?" When we get over that hurdle, then the next group that needs to be convinced is the investment community, and Kathy speaks to them. Kathy needs to be convinced of our results so she can confidently say we're making progress to the investment community. Beyond that, we need to convince the reference labs to buy and start selling our tests. They are convinced when they can think to themselves, "I think we could convince physicians to order this." What really helps there are peer-reviewed publications saying that the markers we're proposing are associated with disease. So we need to start thinking about studies we could do and publish to communicate our findings. Beyond that we need to think about what we could get approved by the FDA, and the product we come up with is not necessarily going to be targeting the largest market, the largest unmet need, but it may be a more narrowly focused product that's more approvable—maybe one that can be proven useful from a very clear-cut clinical trial that's small, short, and inexpensive. Something to get our foot in the door, so to speak.

Chapter Seven

SUMMER 2003

We interviewed James Devlin again on July 24, 2003. As a change of pace, this section will be more summary than direct quotation. Devlin explained some of the technical challenges—such as getting twenty-five different PCR reactions to function in a single tube and then have the amplified product attach to specially prepared beads so they could be read by a laser and genotyped by software. We talked about the details of the coagulation cascade in thrombosis events that is essential to understanding the complex biology of blood clotting. We talked about the SNPs that have been identified to date (and published) as playing a role in this fairly well-understood case—fairly well-understood in that the basic outlines of the cascade have been identified but the functional mechanisms involved have not been adequately characterized.[1] We discussed, in more detail than we had on previous occasions, the multiple levels of gear-shifting required to move from the discovery at Celera Diagnostics of significant SNPs to the creation of a useful diagnostic tool that can be sold to reference labs, to the final and more costly, rigorous, and time-consuming step of preparing a "kit" for the FDA.

Thus, to take a simple example, when a physician makes a determination about the cardiovascular health of a patient, she will consider a small list of weighted risk factors (established national standards). The physician will then look at the patient's lipid levels. If there are two or more risk factors (age, sex, smoking, family history, level of high-density lipids, etc.), as identified by the famous long-term Framingham study, then the physician will order a test for LDL cholesterol. Upon receiving those results, the physician can place the patient into a risk "bin." These bins are quite broad. "What people have a hard time understanding," Devlin explained to us, "is how many gray zones there really are" between these bins. A good genetic test "would help to reduce the gray zones for the physician and make the diagnosis more powerful and hence the treatment as well." The criterion for developing a new test is simple. A genetic test must be both independent of the existing test and increase the accuracy and ease with which a physician assigns a patient to a bin. Consequently, such a test would influence treatment options. It accrues value for the company only if it suggests more aggressive treatment; there are liability issues involved in using a test that might suggest less treatment. Devlin indicated that, because of the liability issues, Celera Diagnostics would not advertise or market the test as a tool for diagnosing when a reduction in the aggressiveness of treatment is appropriate. That means that in all of the diagrams and slides that Devlin uses to map out what physicians could do with the test, arrows always point in the direction of more aggressive treatments. Although the tests Celera Diagnostics is trying to develop take a significantly different approach from those currently in use, they must be linked to the older tests through a simple translation process if they are ever to be adopted widely. The term "translation" is ours.

Receiving FDA approval moves one into a different category. First, it enables a company to sell a kit rather than an ASR (analyte-specific reagent) to a reference lab. Kits can be sold to hospitals and all testing labs and are much more profitable; however, it takes a long time to get them approved, and the costs involved are significantly higher. As we have seen, to convince the FDA, there needs to be a demonstration of biological plausibility for the markers involved. Again, a translation challenge is involved here, though of a different type and requiring different criteria for evidence.

We talked with Devlin about a term that appeared frequently both in our conversations at Celera Diagnostics and beyond, as well as in the published literature: "population." We asked Devlin, "What is a population?" His answer was straightforward and concise, providing us with two different bins, as it were: (1) "One answer is whatever population that we can get our hands on quickly that will allow us to make some reasonable statement, some compelling statement, to a physician in a reference lab as quickly as possible, to start selling something." Understanding the practical needs and constraints of physicians is crucial. The earlier you start interacting with them the better; it is "a very educational process," as it establishes what parameters must be met in the clinical setting for a test to be adopted. (2) A population is any group of people that will provide data. "If possible, I would love to lead with a conclusive publication. But if we can't, I'll lead with whatever I can do quickly within reason. But you would like to have a compelling study to look at risk: Take a population of apparently healthy people and follow them forward in time, and the ones that stay healthy are your comparison group and the ones that get diseased can be studied as your case group. Then you can calculate a relative risk rather than simply an odds ratio." In response to our querying about which such groups are in the nonprospective studies being done at Celera, Devlin replied that in practical terms the population samples that are available are "plain vanilla American or European. So those are the ones we are looking at." There is not much he can do about that, nor did it seem to be a source of concern. Rather, the challenge was to avoid selection bias (such as only studying people who come to a cardiovascular clinic) as much as possible. But even if you can reduce selection bias, normal people still only provide data for a case control study, and in a case control study, you can still only calculate an odds ratio. Thus, it is not as good as a prospective study, which allows you to calculate a relative risk. The obvious downside of prospective studies is that they take a long time and a lot of money. Again, the gold standard of prospective studies is practically impossible to achieve for a company whose goals are short-term, at least under current FDA standards (which are under discussion).

Thus, as of August 2003, the most challenging task is to find a way to translate an odds ratio into a relative risk. As Devlin puts it, "I would like to be able to make relative-risk arguments based on case control rather than prospective studies, but we have so many problems pressing in on us every day that I have not had the time to go there yet. Case control studies are the easiest samples to work with. It is critical that we figure out how to take the odds ratio from those studies and try to apply other outside information, such as prevalence, to convert it into a relative risk. That is what physicians care about. I think that is doable." He is optimistic that the statistical techniques for turning an odds ratio into a relative risk probably already exist in the literature. He points to two textbooks sitting on his desk as well as a stack of photocopied articles neatly arranged above his flat screen computer monitor, above which one sees the photos of his family pinned to the corkboard.

Life Is More Complex: Interview with Isaac Cohen, July 28, 2003

Who is Isaac Cohen? He is an Israeli, a specialist in Chinese medicine who
has a highly regarded practice in Berkeley specializing in breast cancer.
Cohen also has a long-term collaboration with Professor Dale Leitman of
UCSF. Their joint work centers on taking therapeutic herbs from the Chinese repertoire and identifying active molecules in the herbs. Their goal is
FDA approval of new compounds. Cohen and Leitman have a company
named BioNovo. Currently, ten such molecules are in phase-one and phase-
two clinical trials. Hoping for a possible collaboration, Cohen and Leitman
have presented their material to Celera Diagnostics, but without success.

IC: Why do we need diagnostic kits? To help us discover diseases, symptoms, and predispositions, particular problems related to diseases; to
find out what is going on and to intervene, as well as to develop possible
models from which we could study how to prevent those diseases. So
far there is no diagnostic kit that does not have medical intervention/
treatment associated with it. Although there are many diagnostic markers that are irrelevant for the diagnosis of anything specific, like many
blood products, proteins, and genes that are known to us and may even
be associated with pathological conditions, we don't have kits for them.
For example, you can look for women with preeclampsia, which is blood
clotting during pregnancy. That can be very dangerous, but there is no
treatment for it other than to be careful. If you were able to diagnose
that, what would you do? Would you abort? Would you put them on
blood thinners that could be dangerous to the fetus? So there is always
the question, When you build a diagnostic kit, what do you build it for?
And if you had it, what good is it, unless you have something to do with
it that will result in improving the odds for whatever you diagnose? If
every other person will have a thrombo embolic event in their life, what
good does it do to know that they have a three-, or a five-, or tenfold increased risk, unless they knew precisely when, and there was something
for them to do to prevent the onset of it? That something has to be
shown in a prospective trial to reduce the relative and absolute odds of
getting the disease as well as to illustrate that the early detection of the
possible event was a contributing factor to the intervention.

When you study mechanisms of diseases, so far the only way we
have of learning is with snapshots of very small details that may or may

not be related to a whole process, because we cannot see anything in real time. In cancer, as an example even when we are looking at certain times when we arrest cells or stop certain processes it is really artificial to the model we are using; it does not capture the real signaling of cells, an event, that has a certain sequence or a certain way that molecules interact with each other. This brings you to this large question of biological rhythms or biological timing. Biologists don't know timing, we don't even know why embryos develop in the time they develop to become full organisms. And then, as living beings, we have very little understanding of how those events occur so perfectly, with precise timing to elicit this or that. Moreover, if something goes wrong, we don't really know when it occurred. We can talk about things that are related to it—if you had a lot of green tea in your life, it may affect your clotting mechanisms; we don't know exactly why. Everything so far, as far as we can tell, is associative. All we really know is that in the model we are studying, this is what we see as cause and effect. But we cannot really look at living organisms as they are. A perfect example from my field is the timing of surgery in breast cancer and breast cancer outcome. There is quite convincing evidence that surgery done in different times of the menstrual cycle results in significantly different clinical outcomes from the point of view of survival and cancer recurrence. That means that the same cancer develops a completely different phenotype depending on such a temporal thing as the menstrual cycle (twenty-eight days).

PR: There are multiple feedback mechanisms at play?

IC: These systems are constantly re-regulating, so even when it goes awry, it has compensation mechanisms that don't necessarily work in the same linear way that we wish for and describe. For instance, when we are talking about SNPs or mutations or different signaling operations, we don't know when they occurred; if they occurred in a certain time frame. What is their meaning for the population that presents that feature? What does it mean that the Icelandic population or Jewish women have the BRCA mutations in such short evolutionary time? Then you take differences that don't necessarily correspond to a genetic difference. For example, women that live in Asia get five to ten times less breast cancer, unless they come to the United States or Western countries, and then their daughters have the same rate. That means a five to ten rate increase in one generation. Or the previous example of the premenopausal women and their time of surgery. Their gene pool did not change, their SNPs did not change—there is something else going on to permit that kind of aberrant behavior of their breast cells. We also know that genome-wide, epigenetic-like events, like methylation and specific biochemical metabolism like phosphorylation and acetylation of genes and signaling proteins, can cause tumors. So learning about single-point mutations or genes that may be related when you are looking under the microscope may not be relevant to how those diseases display themselves when you are looking at populations in time.

In order to study that, you have to do several things. One of them is that you have to look at the whole population to see if these point mutations are really there, and the other thing is to really understand their mechanisms in relation to how functionally those genes behave. Just because we call the BRCA gene a "breast cancer" gene does not really mean that it is only a breast cancer gene. Now we know that it is also an ovarian cancer gene and may also account for a significant rise in pancreatic, colon, gallbladder, and prostate cancer. Should we call it OVBRCA or a much longer CA? Those are just names. And when people do start looking at functionality, it may be related to some gene for a DNA repair mechanism that is related to some physical stress like radiation. What that may mean is that it predisposes you to be more susceptible to environmental influences that are carcinogenic to anyone. We don't really know what the function is.

PR: So, a mutation in a BRCA gene is probably relevant, but its significance may be unclear?

IC: Well, I am not sure it has relevance. We know that those gene carriers have a penetration to get breast cancer in their lifetime anywhere from 35 percent to sixty percent, even 80 percent. That is a lifetime absolute risk. Now, what does that mean? If you live to eighty-four years old, the total population has a chance of one in eight (13 percent) to encounter breast cancer. Germ line mutations in BRCA1 and 2 and a few other variants account for only 15 to 20 percent of breast cancers that cluster in families and less than 5 percent of breast cancer overall. So having the gene is three- to sixfold worse than the whole population? For those women identified with it, it is much more but yet unclear.

Will we be able to prevent it if we exposed them less to radiation? It is not very different from other risk factors if you make them associative as well. Because it is a gene, we want to believe that it is much more significant. The main difference so far is that we threw much effort to calculate the significance of this risk for women, like the median age for this specific risk when compared to women with breast cancer who are not BRCA carriers. We never calculated it for atypical breast hyperplasia, which seems to be a very significant risk factor for many more women than the few who carry this germ line mutation. The reason is that we don't have a "pretty" scientific story to tell about something so broad as atypical breast hyperplasia.

To get at the significance: Do you take a cohort and see how many of them got it? Or do you look at the general population and see how many of them have it and then see what it really means for the whole population? When you are saying that we are looking only at those with the BRCA mutation, you are already training your observation to be myopic, because very few women have it.

PR: This is very clear. Presumably, now that we have the genome sequence, the range of polymorphisms and mutations will become clearer. What if we did know the distribution of all the genes in *Homo sapiens*, would it be the royal road?

IC: I am not sure. It is possible that it is a question of crunching a lot of genes with a lot of samples. We are not sure. The question is, How come there is variability in different populations according to different diseases? Let's take breast cancer. In very poor countries, like in Africa or parts of Asia, there is very little breast cancer, and even when you do age-adjusted incidence, there is from five times to fifteen times less. When you look at arrays, there is very little difference. You do find various genes that are amplified or deleted in these tumor displays, but those are not breast cancer specific; they are related to things that are already being treated with different therapies, like the estrogen receptor or estrogen dysfunction, different signal transduction genes. If we see these genes, what does it mean in relation to getting or not getting breast cancer? When you look at adjacent tissues, they look pretty much the same. From the point of view of the disease progress, we know that as the disease advances from simple breast hyperplasia to atypical hyperplasia to the in situ stage and forward all the way to metastatic disease, the prognosis gets poorer. You would think that you would observe more genomic instability and more specific patterns that would differentiate this seemingly linear progression of the disease. But what we find is that there is significantly more genomic instability in the in situ phase than in metastatic disease and that the patterns people have identified so far have nothing to do with any genes that are disease specific. One such attempt to characterize the progression is based on cytokeratins, which are present in all cancers, and although the authors of that study named their findings in accordance to the cell origin within the breast duct, they never verified it anatomically. So the current speculations are that the progression beyond the genomic instability is in the stroma, and the whole carcinogenic process starts with a telomeric crisis. Whatever either of these means—they are wild cries in the dark. People still try to force some explanation to a complex picture that is still very unclear. We really lack the stamina to remain in uncertainty. The new church of science needs a dogma, and it manufactures one in advance whenever the previous one is found incomplete. Each lab looks at a small cohort, and they each ask the question in a certain way, and they each get a result that somehow luckily matches that statistical power. Yet, for 90 percent of breast cancer there is no predictive test. We call it spontaneous. No genes, lifestyle, family history can allow us to tell women with any certainty that they will get it or not.

PR: So where you are born—in Africa or New Jersey in the twenty-first century—is a strong predictor?

IC: Yes, if you stay there. In one generation after you immigrate, that would be evened out. Diet, lifestyle plays a role, but what we don't know is what is it precisely that causes the increase in incidence for these immigrants. We believe that the reproductive hormonal milieu plays an important role in breast cancer, mainly because we can treat breast cancer through drugs that target this pathway. Besides, it makes natural sense. Women have significantly different breasts than men, anatomically and functionally, and, surprise! they get much more breast cancer than men. So it must be what makes men and women different, i.e., their sex, reproductive difference. So if it is the reproductive health, innate (inherited), or lifestyle-dependent, when do the changes that result in breast cancer start? In the embryo? In puberty? Childbearing? Menopause? This escapes a monolithic view of the machinery.

There are two ways we try to study this. First is to get a cohort where you get women with a history of breast cancer and look at them. Or to say, "No, no, no, I am just going to take, randomly, a slice of the population, large enough to capture a few of those women who may have this disease, then see if there is any correlation. To do that, in order to get about 150–200 cases in five years, you need to study 15,000 women. That is what was done in the tamoxifen prevention trial, where they enrolled women with high risk, as determined by a multiplier of the known risk factors. In order to do it in one year you have to multiply that not just fivefold but many-, manyfold, because these are rare incidents. Another example from the other direction is the results of the Women's Health Initiative trial, where after five years of treatment with hormone replacement therapy (HRT), we found that we will see 8 additional cases for every 10,000 women using HRT each year. The whole trial had 16,000 women, 8,000 using HRT and 8,000 using placebo, and in five years we saw a total of about 200 cases of breast cancer. Just a little more in the HRT group; little, but significantly more.

Cumulatively, the incidence of breast cancer seems large because we diagnose 200,000 cases a year . . . in a population of 130 million women, of which about 80 million can get breast cancer. So, we say that the lifetime penetration is very high—1 in 8—but spread out so thin that, yes, it is significant for that woman, but when you are studying the big picture, it is very small. So you run a huge risk of not capturing the right thing. When you are doing a trained analysis, you set up a group with breast cancer and another who supposedly don't have it, that you randomly select from your local population. You run the risk of seeing things statistically that may not be relevant once you spread them out. And if even analytically you try to predict, of a hundred genes you think are relevant and you have found fifty SNPs—wonderful!—if you change one, what happens? Or if you add one, what happens? What kind of population data will you need to prove or disprove the importance of the addition or subtraction of one? The sample size increases very dramatically. So you say, okay, I can't really study this, so I will ask instead, What is the sensitivity of the test? This means, What is the percent accuracy that what I am telling you is really what is there? The specificity is that I see it in you. The larger the group and the narrower the number of variables—in our case, genes—you get more accuracy. If you are looking for one thing and you have a large number of people, then you can say if it appears here, then that is enough. But with a large number of variables, or a small group, not related to the general population, then you run the risk that the sensitivity is very low and you have no specificity.

PR: What is a population?

IC: For what question? In our example, if you are trying to predict the incidence and onset of breast cancer, it will be the general public—women between thirty to eighty-four. If it is, for instance, to predict prognosis only, then your population is the women with breast cancer. So the population is really what you define as the relevant population for your study. In science you need to answer a specific question, don't you?

PR: Is it a constructed object?

IC: It can be a constructed object. In a good study it should be a constructed object. It is hard in reality to do it that way. A good population would reflect what you know statistically is happening: risk factors, confounding factors. Including these would help you to do a better population study. But if you add variables, you have more problems, and if you drop variables, you have populations that may not match. Anyway you look at it you lose, because if you want to include what we already know, you have to have a huge sample; if you don't include what we already know, you run the risk that you are not looking at the proper picture. Since we are just taking a snapshot of something, it is really critical that it happens when you say, "Cheese!"—not when you are turning your head.

Predictive kits need to use linear logic: they need to say something that includes this or that as an independent factor that stands alone after you looked at other possible confounding factors tilting it this way or that way. Nature doesn't seem to like linear logic very much. Things are more bell-shaped than linear. Let me give you an example from the diet world. People tried very hard to look if high fat or high protein or body mass index play a linear role in the incidence of breast cancer. There were many conflicting reports. Some concluded, yes, there is a relationship; some said no; while others said that it remained inconclusive. Until a nonlinear analysis of the same studies was done. We found that the extremes of these dietary trends were bad and the center was good. The bell shape was not always a pretty Gaussian distribution, but nonetheless it was definitely not linear. With predictive kits you will have to do major adjustments to define such nonlinearity if you wanted to create a usable scale. Since we don't have a simple pretty picture of genes related to breast cancer, a linear scale is very difficult to find. It can be created, but it may very well be an artificial one.

PR: By definition, scientific objects are constructed. So we are not talking about natural populations out there?

IC: There are natural populations, but you define them in a different way. The appropriate study is not a retrospective study; it is a prospective study. It is a constructed object, but in order to study it, you have to narrow it to what is feasible—sometimes you have to make adjustments. What happens when you look at large gene expression arrays of 80–120 genes? It sounds good, but it is misleading. They have not been compared for other cancers, and hence we don't know the precise relevance for breast cancer. You can number-crunch, but you should do that prospectively. If you want to do it retrospectively, saying, "I know that those women got it, and those women didn't get it," this limits your predictive value, because you have no clue about the timing. All you can say is that this picture tells me that heaven will fall in the next fifty years. Let's say we design the kit, and we want to introduce it. You have to do this huge cohort study to really see the predictive value.

PR: Which they don't have time to do.

IC: Let's pretend we live in an idealistic world. How would it work? Absolute risk is basically irrelevant. So what is relative risk? This is a question you can't really answer without a prospective trial. It took ten or twenty years in the cases where we know the functions—as in an estrogen receptor—to see the relative risk. What does it mean if a woman has this receptor expressed in the tumor or not? To answer that question takes a long time. With the HER-2 gene we had other experiences, and we are in about the fifteenth year that people are doing trials, massive trials, and again you can ask only one question, because there are too many confounding variables. The more you know, the more you have to prove against what is known. In the case of a genome-based analysis, all we know is the descriptive character and nothing else related to it.

I think it is great to study SNPs, but it is not necessarily the next step to create diagnostic kits that just have them. Knowing the pathway lets you know if it is relevant when you have a treatment. The kit gives you an absolute risk. What do I do as a physician? Okay. Someone comes to me as a physician and says to me, "I have a thirtyfold risk." Let me tell you what that means: You are now forty-five, so until you are eighty-five you have a thirty times greater risk of getting breast cancer. That means instead of twenty-five women out of a thousand, you will be higher. When you add other risk factors, you still have only two options nowadays. One is yank off your breasts and hope the cancer hasn't started, and the other is to take some drug that may or may not be relevant. The third is to say, "I am going to live my life." From what I understood about what they are doing about breast cancer, Celera is not going to get at the relative risk—unless they run a trial that proves the principle that their kit is predictive over time.

PR: Devlin said exactly that at the end of the last interview: "We have odds ratios and relative risk. What we would really like is to find a means to translate one to the other."

IC: These are two very different things. There is no reconciliation.

PR: Why?

IC: You can't look to the right and say what is happening on the left. They will never get even the proper odds ratios, because of the problems we have been talking about. In the WHI study with 16,000 women, looking at one treatment against placebo, prospectively, they were trying to establish the hazard ratio, and they wanted to do it in a ten-year time frame but ended up doing it in a five-year time frame because there was more breast cancer. If you look carefully at those numbers and ask, How many events actually happened? About 140 breast cancers versus 110 in the placebo. Is that significant? So you do the hazard ratio, but it is false, because the women in the trial were willing to take hormone therapy while the majority of women out there are not. This small difference is meaningless to the general population but meaningful for the women willing to take HRT. This is when you are studying a therapy. When you are studying a diagnostic predictor kit, you are running into much greater risk of misconstruing the significance of the result.

PR: They would answer that there is an identifiable difference between those two groups.

IC: This is a wish, but they don't have it. It is everybody's logic, but they don't have it. We said the question is why this small group makes it impossible for a very small group to prevent others from taking a good drug. Can we identify the very small group and exclude it? At 60 or 80 or 100 percent confidence? The pharmacogenomics may have to do with the drug. You have to study the drug from this point of view and study the population. Those blood samples for that population already exist; we have them. Then we could relate it to what that drug does, which is functional genomics, not SNPs. This is a much smaller but more precise question. This is our question. We want to predict something that is a much smaller market. Are women at a high risk or low risk for this drug? The next question is, Okay, if we discover that, can we extrapolate from that who might be at risk for this disease or that disease, et cetera? That doesn't interest them. I wouldn't use diagnostic predictor kits as a physician today. They are not strong enough.

The View from Affymetrix:
Interview with Steve Fodor, September 26, 2003

Steve Fodor is founder, chairman, and chief executive officer of Affymetrix, a leading player in the gene chip market. Fodor received his Ph.D. in chemistry from Princeton University. Between 1986 and 1989, Fodor was the National Institutes of Health postdoctoral fellow at UC Berkeley. He was also a pioneer in the discovery and development of microarray-based genomic analysis. Affymetrix primarily sells tools for studies in genetics. While Fodor observes the discovery strategies of other companies in the areas of molecular diagnostics and therapeutics, he himself is involved in providing such efforts, both in academic and nonacademic settings, with the technology to proceed. Celera, however, is not using the gene chip technology.

We spoke with Fodor at Affymetrix.

PR: Although our agreement was that you wouldn't comment directly on other companies, could you give us an overview reaction to the strategy of focusing on functional SNPs?

SF: I think there's a pretty good consensus now that the protein coding content of the human genome takes up around 1½ to 2 percent of the nonrepetitive segments of the sequence. And I think that it is inherently pleasing to think about focusing on that portion for function. It's because of our history of success with some single-gene disorders. There are many documented examples of single-gene disorders altering protein structure and altering function. But monogenic conditions are just a small part of the pie. The reason monogenic diseases have been discovered in the genome is because they're apparently relatively recent mutations that are catastrophic, so you can observe the effects very easily. You can look in families; you can do linkage measurements and so on, or actually target the region in the genome easily.

When you start to look at polygenic diseases, the question is, If ten genes are involved in causing a disease, is it that each of those has been slightly varied structurally so that each one acts a little differently? Or is there a difference in the amplitude of these genes and in the expression of these genes? For example, for most of our physical characteristics, the function of the genes is the same in all of us. And yet slight changes in the amount of expression will cause my ears to be bigger than yours. When we think about complex metabolic pathways, for example, in diabetes or cardiovascular disease, if we find functional changes in the structural components of these genes, we don't know if they are going to act in aggregate to cause these effects, or are they going to be more subtle effects that are dispersed in the genome, particularly in regulation of these genes? So I would argue that there's a whole lot more to look at than just SNPs in regions of the genome that code for proteins. We now know that while the coding region comprises about 1½, 2 percent of the sequence, there are regulatory regions for all these genes. And they might be transacting with all of the parts of the genome. We now know that there are hundreds of thousands, if not millions probably, of different RNA molecules that are swimming around in the cell and involved in the regulation of all of the genes. There's a whole intracellular communication going on. I believe that variation of the genetic sequence within the regulatory region and other mechanisms that we do not yet understand are going to be important in creating a good description of the genome.

PR: There's a strong version and a weaker version of what you're saying. The strong version would be: It's premature to be moving towards definitive claims on which to found diagnostic tests because we're only beginning to understand how the genome works. The weaker version is that we're in a revolutionary period, we're learning a lot, and so on. Hence, there will always be more to learn.

SF: You're right. Clearly, there's a long-term ambition to try to unravel the secrets of the genome, and I think it is vital to build tools that allow you to do that. In 1994, at Affymetrix, chips for expression monitoring 10,000 genes were pretty expensive. They were probably a couple of thousand dollars apiece. Today, and by the time your publication comes out, academics can buy a chip that looks at 61,000 transcripts for around $300. Technology is technology. In the beginning, of course, there's very short supply, and there's a tremendous amount of R&D that goes into a product, so it tends to be highly priced. And now, in 2003, we've got multiple types of chips that actually start to screen the entire genome, not just this 1½–2 percent, with very, very high resolution. And again, are those expensive today? Well, yeah, probably. On the other hand, it doesn't cost you $2 billion to resequence the genome anymore. You can now do it for thousands. Over time, that's going to come down, too.

But back to your point: You could make the argument that you can wait to use these technologies until you really mechanistically understand phenotypes and so on, but from a diagnostic perspective, think what people do today with MRI. A physician goes and puts the patient in and then gets an image. And so a physician sits down in front of the computer or gets the pictures and goes through them and says, "Aha, here is something I want." Well, is the MRI telling the physician what is happening mechanistically? No. What it's doing is alerting the physician to something that has happened.

I think there are a couple of big principles. One is, in some senses, we think of processes in biology the same way we think of an event such as the fan belt breaking in your car and the car stopping. What do you do? Just put a new fan belt in. Biology doesn't work that way. You tweak one thing and fifteen other systems compensate, and that's true for every ecological or biological or physical system we know.

PR: Or social system.

SF: Or social—everything, right. And yet—and this is my primary objection to limiting yourself to the coding sequences of the genome—it is a system that is composed of multiple components and regulatory elements that we do not yet understand. Again, do I believe there are fan belts that can break? You bet. Do I believe it's the whole story? Absolutely not. When we start to look at these models of the genome, we treat it like a mechanical system. The genome, we don't even know what language it's been written in! So it's unlikely that the same approaches, when we think about it—that involve linear systems, are going to be applicable to the genome. I think that, long-term, what we want to do is to understand the underlying biology behind all these things. At the same time, there's enormous value in providing technology that gives people guidance in an area where they have very little guidance at this point.

Highly Probable Success:
Interview with Mike Hunkapiller, July 7, 2003

PR: We gather it's crunch time now.
MH: Yeah, it's crunch time, that's right [*laughs*]. It's day to day. I think, so far, in every one of those disease areas that they've looked at, the technology application seems to be working very well. As you would hope, in a sense they got some surprising results in some of these areas in terms of what the associations were. There were even questions early on: Does that make sense? When they went back and really looked at the science that was known about those genes, they thought, "Yeah, it does make sense," and it seems to be working really well at this point. The genes they found were not necessarily ones that people had already associated with the specific disease in question. CDx seem to have gotten results that in retrospect, when you went back and looked at what was in the scientific literature, made biological sense.
PR: We are now into a period in which the Celera Diagnostics machinery is operational. We know the meetings are taking place in specific disease areas. There is a lot of stress and a lot of tension. In your opinion, is it inevitable that it's going to succeed?
MH: Well, nothing is ever a sure thing from a business perspective until you have sales in hand, but I think the answer is yes. I think the technique

certainly works. Will it work in every disease study that they do? It probably won't. But it doesn't have to; they've done a very careful job of grading the opportunities from a business perspective before making choices as to what diseases to go after initially. I think the chances of success are very high. I've got no reason to believe that the general approach is wrong. It seems to be, as I said, even more successful on individual experiments than they would have predicted it would have been. You know, there are all kinds of land mines out there, from intellectual property issues, to the priorities of how the healthcare providers spend their money, to approvals by government agencies. Those are semiquantifiable risks, and you reduce those by getting people that understand them and plan accordingly. I can't imagine a better group of people to be able to do that wisely. Well, I consider persuading Tom, John, and Kathy to join as one of my better accomplishments here in twenty years [*laughs*].

❏

Iain Pears's *An Instance of the Fingerpost*, an ingenious mystery set in seventeenth-century England, tells its story from multiple viewpoints that shift the time frame from prospective to retrospective and back to a fictionalized present. As a means of rhetorically establishing his virtuous character, so that readers will believe his account of events rather than the divergent accounts that have preceded it, one of the characters, an imagined version of a real Oxford historian, proclaims, "Besides, my job as an historian is to present the truth, and to tell the tale of these days in the approved fashion—first causes, narrative, summation, moral [. . .]."[2]

The reader of our chronicle will have long since surmised that no such rhetoric is employed here. We have established our authority mainly through the willingness of those encountered in these pages to talk to us and to allow us to craft their accounts. We have made the scantest effort to establish our virtue, but, as this is no longer the seventeenth century, this neglect is of no consequence. The account's veracity will be evaluated on other grounds. Readers would surely be distressed at this point to encounter a discussion of first causes from characters claiming to be anthropologists of the contemporary—and our intent is surely not to distress our readers unduly. Our chronicle has no moral, although throughout it does treat practices of ethics, understood as self-formation.

But while we felt no need to prove our authorial virtue, or to treat of first causes, or to draw a moral, we did sometimes wonder, as we ferried back and forth on highway 880 between Berkeley and Alameda, if our narrative required some kind of "summation," the fourth component of the traditional history. Did we need to provide one type or another of summation? Dan-Cohen, displaying her consistency of mind and character, on a number of occasions reminded Rabinow, when his will weakened, of the genre constraints of the chronicle. "Besides," we might have said, "our job as anthropological chroniclers of the contemporary is to tell the tale of these days in an appropriate fashion—find a form for the natives' points of view, leave things in an appropriate state of irresolution, provide sufficient material for readers to find their way in a landscape where representations of practices of truth, ethics, and power relations are made available for reflection through a studied figuration." But, had we said that, most of our readers would long since have abandoned us—and we seek your company. What we can say is that we pondered with a mixture of pleasure and perplexity the not entirely discordant juxtaposition of truth claims that had been presented to us in the month we designated as the last one of fieldwork. This unsettled state of affairs, opinions, and methods, after all, provided a certain validation for our choice of narrative form.

Illustrations

Joe Catanese

Gabriella Dalisay

James Devlin

Shirley Kwok

Victor Lee

Kathy Ordoñez

John Sninsky

Tom White

In 1985, Cetus Corporation, a biotech start-up company located in Emeryville, California, and Perkin-Elmer Corporation, an instrument company located in Norwalk, Connecticut, formed Perkin-Elmer Cetus Instruments (PECI), a joint venture to develop new instruments such as polymerase chain reaction (PCR) thermocyclers. Hoffmann La Roche is a Swiss pharmaceutical company with major operations in the United States. In 1989, Hoffmann La Roche and Cetus began a five-year research and development collaboration to commercialize diagnostic applications of the polymerase chain reaction invented at Cetus. In July 1991, a new subsidiary of Hoffmann La Roche was established, Roche Molecular Systems. In December 1991, Hoffmann La Roche acquired the rights for the core PCR patents developed at Cetus. Cetus ceased to exist. The patents for PCR instrumentation were acquired by Perkin-Elmer. Several of the key actors in this book—Kathy Ordoñez, Tom White, John Sninsky—worked in high management positions at Roche Molecular Systems and at Cetus. Interested readers can find more detail about these events in *Making PCR: A Story of Biotechnology*, by Paul Rabinow.

In 1993, Perkin-Elmer acquired another instrument company, Applied Biosystems in Foster City, California. The executive vice president of Applied Biosystems was Michael Hunkapiller.

In 1995, Tony White (no relation to Tom White) was named president of Perkin-Elmer.

In 1997, Hunkapiller was named president of PE Biosystems (later Applied Biosystems Group).

In 1998, Celera Genomics was founded as a new business unit of Perkin-Elmer. Its president from 1998 to 2002 was Craig Venter.

In 1999, Perkin-Elmer sold part of its instruments business, along with the PE name, and changed its name to Applera Corporation. Celera Diagnostics (CDx) was formed in April 2001 to commercialize the diagnostic applications of Applera's technologies.

In January 2002, Craig Venter left Celera Genomics. He was replaced in April 2002 by Kathy Ordoñez, who has also been president of the new company, Celera Diagnostics, since its foundation.

Today, Applera is composed of three business entities: Celera Genomics, Celera Diagnostics, and Applied Biosystems.

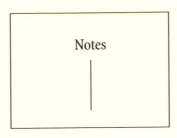

Notes

OVERTURE
A MACHINE TO MAKE A FUTURE

1. Our heartfelt thanks for the insights and support of Roger Brent, Carlo Caduff, Stephen Collier, James Faubion, Nicolas Langlitz, Stephanie Pifer, Peter Skafish, Janet Vanides. Our gratitude to all those who agreed to work with us—generously, patiently, tolerantly.

2. Sydney Brenner, "The End of the Beginning," Drosophilia issue of *Science*, no. 287 (March 24, 2000 [*Drosophilia* issue]): 2173–74.

3. François Jacob, "Time and the Invention of the Future," in *The Possible and the Actual* (Seattle: University of Washington Press, 1982), p. 16. Cited in Hans-Jorg Rheinberger, *Toward a History of Epistemic Things: Synthesizing Proteins in the Test Tube* (Stanford: Stanford University Press, 1997), p. 182.

4. François Jacob, *The Statue Within: An Autobiography*, trans. Franklin Philip (New York: Basic Books, 1988), p. 234. Orig. *La statue intérieure* (Paris: Editions Odile Jacob, 1987). Rheinberger, p. 25.

5. Rheinberger, p. 28; Jacob (1988), p. 9.

6. Bruno Latour, *Science in Action: How to Follow Scientists in Society* (Cambridge: Harvard University Press, 1987). Simon Schaeffer, "Late Victorian Metrology and Its Instrumentation: A Manufactory of Ohms," in *The Science Studies Reader*, ed. Mario Biagtoli (New York: Routledge, 1999), 457–478.

7. The phrase "from the native's point of view," is from one of the founding texts of modern anthropology, Bronislaw Malinowski's *Argonauts of the Western Pacific* (1922; New York: E.P. Dutton, 1961), p. 25.

8. More details in the December 2003, *California Monthly*.

9. Niklas Luhmann, "European Rationality," in *Observations on Modernity*, trans. William Whobrey (Stanford: Stanford University Press, 1998), p. 35.

10. For more information on this topic, see the following sources: Kevin Davies, *Cracking the Genome: Inside the Race to Unlock Human DNA* (New York: Free Press, 2001); Nicholas Wade, *Life Script: How the Human Genome Discoveries Will Transform Medicine and Enhance Your Health* (New York: Touchstone Books, 2001).

11. Written testimony of Gerald M. Rubin before the Subcommittee on Energy and Environment of the House Committee on Science, April 6, 2000. Available online at http://clinton4.nara.gov/WH/EOP/OSTP/html/00626_5.html.

12. Ibid.

13. Wade, p. 62.

14. "Celera Primes for Key Change," *Baltimore Sun*, April 28, 2002. Available online at http://www.washingtonpost.com/ac2/wp-dyn/A53957-2002Aug 7.

15. Ibid.

16. Keith Epstein, "Can She Do It?" *Washington Techway*, August 7, 2002.

17. "Celera Primes for Key Change."

CHAPTER ONE
EXIT AND ENTRY

1. Albert Hirschman, *Exit, Voice, and Loyalty: Responses to Decline in Firms, Organizations, and States* (Cambridge: Harvard University Press, 1970).

CHAPTER THREE
THE MACHINERY AND ITS STEWARDS

1. Steven Shapin, *A Social History of Truth, Civility and Science in Seventeenth-Century England* (Chicago: University of Chicago Press, 1994), p. 356.

2. Ibid., p. 358.

3. Ibid., p. 361.

4. Ibid., p. 414.

5. Paul Rabinow, *French Modern: Norms and Forms of the Social Environment* (Chicago: University of Chicago Press, 1989).

6. http://boisechurchofchrist.org/SermonsCox/Stewardship.htm.

CHAPTER FOUR
ETHICAL AND CULTURAL CONSULTANCY

1. Niklas Luhmann, "Familiarity, Confidence, Trust: Problems and Alternatives," in Diego Gambetta, ed., *Trust: Making and Breaking Cooperative Relations* (Oxford: Blackwell, 1988).

2. Ibid., p. 99.

3. Ibid., p. 100.

4. Michael Mulkay, quoted in Paul Rabinow, *Making PCR: A Story of Biotechnology* (Chicago: University of Chicago Press, 1996), p. 12.

5. For a recent survey, see: David Blumenthal et al., "Data Withholding in Academic Genetics: Evidence from a National Survey," *Journal of the American Medical Association* 287, no. 4: 473–80 (2002).

CHAPTER SIX
MODELS ORIENT, TECHNOLOGIES PERFORM,
SAMPLES SPEAK (OR VICE VERSA)

1. One could draw parallels with the emergence of other specialties, such as biotechnology patent law: molecular biologists were sent to law school by their companies or firms to study patent law and to invent the interface with the emergent discoveries. Only later was sufficient case law developed to have courses taught on the subject and specialists earn their stripes.

2. Quanhe Yang, Muin Khoury, J. Friedman, and W. Dana Flanders, "Improving the Prediction of Complex Diseases by Testing for Multiple Disease-Susceptibility Genes," *American Journal of Human Genetics* 72 (2003): 636.

3. Ibid.

4. Ibid.

5. Ibid., p. 644.

6. Ibid.

7. Ibid., p. 645.

CHAPTER SEVEN
SUMMER 2003

1. See, for example, Ida Martinelli, "Risk Factors in Venous Thromboembolism," *Thromb Haemost* 86 (2001): 395–403; Uri Seligsohn and Aharon Lubetsky, "Genetic Susceptibility to Venous Thrombosis," *New England Journal of Medicine* 344, no. 16 (2001): 1222–31; Juan Carlos Soulto, John Blangero, et al., "Genetic Susceptibility to Thrombosis and Its Relationship to Physiological Risk Factors: The GAIT Study," *American Journal of Human Genetics* 67 (2000): 1452–59; Jan Vandenbroucke, Frits Rosendaal, et al., "Oral Contraceptives and the Risk of Venous Thrombosis," *New England Journal of Medicine* 344, no. 20 (2001): 1527–35.

2. Ian Pears, *An Instance of the Fingerpost* (New York: Random House, 1998), p. 561.

allele	One of several possible forms of a gene, usually defined by an association with a specific phenotypic trait.
amino acids	The building blocks of proteins. Each amino acid is coded for by a combination of three nucleotides.
assay	A test that measures the functions or properties of a mixture or compound.
base*	A unit of the DNA. There are four bases: adenine (A), guanine (G), thymine (T), and cytosine (C). The sequence of bases (for example, CAG) is the genetic code.
base pair*	Two DNA bases complementary to one another (A and T, or G and C) that join the complementary strands of DNA to form the double helix characteristic of DNA.
bioinformatics	The analysis and organization of biological data.
candidate gene	A gene that is thought to contain markers that are linked to a specific disease.
cDNA*	Complementary DNA: single-stranded DNA made in the laboratory from a messenger RNA template under the aegis of the enzyme reverse transcriptase. This form of DNA is often used as a probe in the physical mapping of a chromosome.
cDNA library*	A collection of DNA sequences generated from mRNA sequences. This type of library contains only DNA that codes for proteins and does not include any noncoding DNA.
chromosome	A biological structure in the nucleus of cells that is composed of DNA and contains many genes.

*Definitions taken from www.medterms.com.

coding SNP	A SNP that occurs in a region of DNA that codes for the production of proteins.
deletion	The removal of one or several nucleotides from a segment of DNA.
disease association studies	Studies aimed at identifying genetic markers that are linked to a particular disease.
DNA probe	A single-stranded DNA molecule used in laboratory experiments to detect the presence of a complementary sequence in a mixture of other singled-stranded DNA molecules.
DNA sequencing	Determining the order of nucleotides in a segment of DNA.
enzyme	A protein that can be used to speed up or slow down a reaction, while itself remaining unchanged.
expressed sequence tag (EST)*	A unique stretch of DNA within a coding region of a gene that is useful for identifying full-length genes and serves as a landmark for mapping. An EST is a sequence tagged site (STS) derived from cDNA.
expression array*	A way of analyzing the differential expression of thousands of species of mRNA at the same time in two different samples (as, for example, in the normal vs. tumor tissue, or at different developmental stages).
functional genomics	A field of study that concentrates on the function of genes in organisms.
gene**	The basic unit of heredity; the sequence of DNA that encodes all the information to make a protein. Structurally, a gene is formed by three regions: a regulatory region called the promoter, juxtaposed to the coding region containing the protein sequence, and a "3' tail" sequence. In mammalian cells, the promoter is a complex region containing binding sites for many proteins that regulate gene expression. A gene may be "activated" or "switched on" to make protein—this activation is referred to as gene expression—by proteins which control when, where, and how much protein is expressed from the gene. In the human genome, there are an estimated 28,000 genes. Some of these are evolutionarily related

**Definitions taken from Celera's web glossary.

and form "gene families" that express related proteins. There are also genes that no longer make a protein; these defective remnants of evolution are called pseudogenes.

gene expression
The forming of an RNA or protein product from the information contained in a gene.

gene expression profiling
The study and determination of differences in the expression levels of genes.

genetic markers
DNA polymorphisms that act as experimental probes and can be used to locate genes.

genotype**
The genetic constitution (the genome) of a cell, an individual or an organism. The genotype is distinct from its expressed features, or phenotypes. The genotype of a person is her or his genetic makeup. It can pertain to all genes or to a specific gene.

heterozygous
Having two different alleles for the same gene, each of which is inherited separately.

homology
The relatedness of biological entities in evolutionary terms.

homozygous
Having identical alleles for a gene.

host-response studies
Studies that probe how a particular organism will respond to specific treatments.

insertion
The addition of one or several nucleotides to a segment of DNA.

interferon
An antiviral protein produced as a biological response to an invading virus. Interferon can be used to treat some diseases.

linkage region
A region of DNA that is characterized in terms of the likelihood that the genes in the region will be inherited together. Genes that are closer to each other on a chromosome are more likely to be inherited together than ones that are far from each other.

medicinal chemistry
The use of chemistry in researching, developing, and designing new drugs.

messenger RNA (mRNA)
A molecule of ribonucleic acid that transmits information from the DNA in the nucleus to the cytoplasm, where the process of protein synthesis starts.

molecular diagnostics
The identification of disease or disease risk through the analysis of DNA or RNA.

monogenic
An inherited disease or trait caused by a single gene.

mutation	A biochemical alteration in the DNA sequence.
nucleotide	The basic unit of DNA and RNA, composed of a sugar, a phosphate, and a base.
penetrance**	The likelihood a given gene will result in disease. For example, if half of the people with the neurofibromatosis (NF) gene have the disease NF, the penetrance of the NF gene is 0.5.
pharmacogenomics	A field of study which aims to find the correlation between genotypes and the responses of specific patients to drug therapies.
phenotype	A biological characteristic resulting from the expression of a gene in an organism.
polygenic	An inherited disease or trait caused by multiple genes.
polymerase chain reaction (PCR)	A method used to amplify segments of DNA through repeated replication.
polymorphism	A naturally occurring variation in the DNA sequence.
Positional cloning**	Cloning a gene based simply on knowing its position in the genome without any idea of the function of that gene. Because this is the reverse of how things have been traditionally done, it has also been called reverse genetics.
primer	A short chain of nucleotides to which DNA can be added.
prospective study**	A study in which the subjects are identified and then followed forward in time.
protein	A molecule composed of amino acids. The order of the amino acids is encoded in cellular DNA.
proteomics	The study of protein structure, composition, and function.
reagent	A substance that is used to produce chemical reactions.
regulatory SNP	A SNP found in a region of DNA that regulates the carrying out of processes encoded in genes.
resequencing	The process of sequencing DNA from multiple individuals in order to understand the scope of the genetic variations between them.
retrospective study**	In medicine, a disease study that looks backward in time, usually using medical records and interviews with patients already known to have the disease.

reverse transcription	The process by which a complementary DNA molecule is produced from an RNA template.
single-nucleotide polymorphism (SNP)	A naturally occurring variation in one of the four bases that compose DNA.
splicing	A process that takes place after transcription in which portions of the DNA are discarded, the DNA around these portions is spliced together, and the resulting DNA is then used to form the mRNA.
toxicology	The study of undesirable effects caused by drugs.
transcription	The transmission of the information from a DNA sequence to an mRNA molecule in the creation of mRNA.
translation	The conversion of the information in an mRNA molecule into a corresponding sequence of amino acids in the construction of a protein.